Star Stories

Star Stories

Constellations and People

ANTHONY AVENI

Yale UNIVERSITY PRESS

New Haven and London

Published with assistance from the foundation established
in memory of Philip Hamilton McMillan of the Class of 1894,
Yale College.

Yale University Press books may be purchased in quantity for
educational, business, or promotional use. For information, please
e-mail sales.press@yale.edu (U.S. office) or sales@yaleup.co.uk
(U.K. office).

Set in Minion type by Integrated Publishing Solutions.
Printed in the United States of America.

Library of Congress Control Number: 2019936113
ISBN 978-0-300-24128-0 (hardcover : alk. paper)

A catalogue record for this book is available from the British
Library.

This paper meets the requirements of ANSI/NISO Z39.48–1992
(Permanence of Paper).

10 9 8 7 6 5 4 3 2 1

To

Robert H. N. Ho, whose gift to Colgate University
of the Ho Tung Visualization Laboratory inspired me
to share its sky imagery with a wider audience

Contents

Preface

Before smartphones, we had books; before books, we had images on cave walls or in sacred temples, their messages enhanced by spoken words now lost. The sky, too, has long been a canvas for telling stories about the meaning of life. Early humans looked for hidden likenesses between the domains of heaven and earth—hoping, with the celestial imagery they envisioned, to marry the unfamiliar above to their quotidian lives below. Our contact with the sky humanized us. It encouraged us to use our imaginations to tell stories about who we are.

Like nothing else in the natural world, the heavens were pristine, perfect—the ideal place for the gods to reside. Celestial time rolled on in endless cycles, portending our fate. What better way to peek around time's corner into the future? What could be a better medium for creating tales with moral significance than the silent, dependable courses of the stars, coming and going with the seasons—constellations that heralded births and deaths, reminded us of times of war and prosperity, and memorialized our personal loves and adventures?

Star Stories focuses on the cultural diversity inherent in cosmic storytelling. Constellations and star groups, conceived by a host of ancient and contemporary cultures, will set the stage for a deep discussion of how nature (climate, environment, latitude) and cultures (from hunter-gatherers to empires) have inspired humans to create a wide variety of narratives using patterns in the sky. These stories, from countless generations that came before, are also now ours to ponder and share.

Star Stories

Patterns

Can you remember lying on the grass on a hot midsummer day? You may have gazed at puffy cumulus clouds roiling in a deep blue sky and imagined familiar forms morphing from one to another: there's a racing car, a baseball outfielder's glove, your dog's face. We do the same with geological formations: New Hampshire's Old Man of the Mountain, Britain's Queen Victoria's Rock, Lot's Wife in Israel, the Woman of Mali in Guinea, and dozens of sleeping giants and giantesses—even the shadowy likeness of an extraterrestrial face in images of the surface of Mars. Our brains are pattern-recognition mavens. Psychologists call this human knack for visualizing unity among random data pareidolia. The mind tries to resolve the tension that accompanies randomness by attempting to perceive something familiar in an otherwise unfamiliar pattern. These apparitions often exhibit religious overtones: Muhammad in a flame, or Jesus Christ on a tortilla.

A thirty-thousand-year-old painting on the wall of Chauvet Cave in France shows a pair of horned auroch, ancestors of our domesticated cattle, their heads lowered and shoulder muscles flexed. Antlered animals sketched in the background watch as the bulls poise for attack. The cave's painter rendered this dramatic tableau with the delicacy of a contemporary artist. I can imagine an extended family sitting cross-legged by a fire staring at the painting, one of them standing close to it, spear in hand; another clad in the skin of an animal feigning an attack. This is the age-old scene of hunter and hunted, an event vital to the subsistence of the group and to the propagation of their lineage. Were they acting in anticipation of what would take place the next day? And was the ritual of enactment required to make the hunt happen? We will never know.

Once the cumulus clouds evaporate in evening twilight and a pitch-black, star-studded sky replaces the azure screen, another backdrop, equally suitable for expressing the narrative of the hunt, appears outside and above the entrance to Chauvet Cave. Stars file across the firmament in a dark sky we experience all too rarely today, thanks to artificial lighting. Shepherds of the ancient Middle East with little more to do than tend their flocks mused about the resemblance of the Big Dipper to a wagon and Orion to a man. The night sky became their natural storyboard, available free of charge to everyone. Since long before our electronic devices, picture books like this one, and even cave walls awaiting the artist's brush, the night sky has been a medium filled with countless points of light that beckon gazers to connect the dots.

Recognizing and naming patterns in the sky gradually became more than just a casual affair: it grew to be part of a deliberate recall of imagery that possessed religious or mythic significance, a reminder of the glory of the gods we praise for creating the world, or of the power of the ruler who proclaimed descent from them. It may have started when priests who, looking upward as they worshipped their heavenly gods, conceived of figures made out of star patterns through which to better express themselves.

To judge by their names, the constellations familiar to stargazers with a Western cultural orientation descend from third-millennium BCE Sumerian civilization. They make their first concrete appearance on boundary stones and cuneiform tablets in the seventh century BCE as well as in the contemporary epic Greek tales of Homer and Hesiod. (Throughout this book, I will often contrast world cultures through history with "the West." By that I mean European-American Western civilization descended from beliefs and customs of the ancient Middle East through the classical Greco-Roman world. The route of descent into the modern, Western world passed through Islam, medieval and Renaissance Europe, and the French Enlightenment.)

Ptolemy, a second-century Alexandrian astronomer, listed forty-eight constellations. Nearly three dozen were named after land animals, fish, and birds, with a sprinkling of serpents and humanoids—as well as one insect. A dozen were added in 1603, when the German lawyer-cum-cartographer Johann Bayer made the first sky map of the Southern Hemisphere. In 1922 the International Astronomical Union put the official list at eighty-eight. It includes the eighteenth-century Enlightenment's tribute to scientific achievement, with constellations picturing a telescope, microscope, air pump, alchemical furnace, and architect's chisel and triangle. Medieval constellations, such as Saint Peter's Keys to Heaven's Gate, were dropped.

China boasts 283 constellations, with names quite different from those with Sumerian origins; the earliest appear inscribed on oracle bones of the Shang dynasty around the fourteenth century BCE. The *Rig Veda*, a Hindu hymn dating to the second millennium BCE, also refers to the constellations, as does the writing in royal tombs of sixteenth-century BCE Upper Egypt. In the Americas, the Navajo, Iroquois, Maya, Incas, and Aztecs fashioned star patterns related to matters of great importance to them. So did the Aboriginal people of Australia, as well as dwellers in the tropical rainforests of South America, the icy landscapes of Arctic Siberia and Alaska, and the deserts, forests, and veld of Africa.

Star Stories is about something we all have in common. We created constellations for discourse about moral issues and social rules, about affairs both practical and spiritual, about our immediate needs and our wildest dreams. Let the retelling of these stories serve as a celebration of the limitless imagination of our human family.

1

Orion's Many Faces

To the Greeks, Orion was a demigod, a son of the sea god, Poseidon. Able to walk on water, Orion stepped across the sea to visit the court of an Aegean island king, where he drank too much wine and assaulted the royal princess. As punishment, the king blinded him, took away his water-walking power, and sent him packing. Rescuing him from this low point was the benevolent fire god, Hephaestus, who took pity on Orion and offered him a servant, Cedalion, to guide him to the place where the sun rose. When Orion and Cedalion arrived at the horizon, Apollo cast his healing rays on the demigod, restoring his vision and sea-walking ability.

Ultimately Orion found refuge on Crete, where he became a celebrated game hunter skilled at the bow. But excess and adrenaline from the hunt got Orion into trouble once again. Orion, now under the tutelage of Artemis, goddess of the wild, boasted that he would slay every animal on the face of the earth, which understandably alarmed Gaia, the earth goddess. Some say she delivered out of her bowels a scorpion, which dispatched the bold hunter by stinging him on the heel.

Others say Artemis herself set the venomous animal on Orion to halt his infamous amorous advances.

The story of Orion is an ancient Greek reminder that anyone guilty of hubris, in this case boasting of divine prowess, will spark nemesis, or retribution, from the gods. This is why Orion, the hunter, appears opposite Scorpius, the scorpion, in the nighttime sky. The starscape offers a backdrop for other salient points in the story as well. Key mileposts along Orion's seasonal journey include the blinding of Orion and his sinking into the sea (which coincides with his namesake constellation's vanishing after sunset in late spring), and the regaining of his vision (when he returns to the night sky in the middle of summer). Orion the Hunter is also most prominent in the sky during late autumn, the season when thoughts turn to the hunt.

Orion the Hunter was once called Al Jabbar, the Giant. In fact, most of the Western constellation names are Arabic. Bright-red Betelgeuse, or Ibt al Jauzah, is the "armpit of the central one" (or less frequently, the giant's shoulder, arm, or right hand). Rijl Jauzah al Yusra, or Rigel, is the bright blue star that marks Orion's left leg. The closely gathered line of three bluish stars of Orion's Belt were the golden nuggets that lay at the middle of the constellation. Each had its own designation. Mintaka, on the right (west), means "belt," while Alnilam (in the middle) is the string of pearls set at the center of the belt. Last to rise, Alnitak is the girdle. Up in the other shoulder lies Bellatrix, the only prominent star in Orion that lacks an Arab designation (it means "female warrior" in Latin), though old maps label this star Al Murzim, or Mirzam, the "roaring conqueror." Less-luminous Saiph marks Orion's fainter right leg. Actually *Saiph* means "sword" and was originally

The Greek constellation of Orion.
(Samuel Lee, via Wikimedia Commons)

intended to mark the tip of the weapon that hangs from the Hunter's belt. The brightest star in the faint and gauzy handle of the sword, which houses Orion's Great Nebula, is Na'ir al Saif, "brightest one of the sword." Al-Maisan, or Meissa, apparently named as the result of an erroneous juxtaposition with a star in neighboring Gemini, was once the Head of Al Jauzah, or Al Ras al Jauzah; and finally, the easily recognizable chain of faint stars along the upheld arm above the right shoulder that represents the sleeve of the garment he wears was called Al Kumm, or "sleeve."

Why these particular stars, why there, in that sky place, visible in that particular season? Who narrates tales of Orion? Who are the listeners? And what do the answers to these questions tell us about the people who tell star stories?

From ancient Chinese dynastic records dated to the twelfth century BCE, we can see that the Orion story was once used for political propaganda. Ebo and Shichen were the first and second sons of the great mythical Emperor Ku, inventor of musical instruments and composer of songs, who traveled his vast empire by horse in autumn and winter and on a dragon in spring and summer. Despite the emperor's great skills at governance, he was burdened with the decision about who would someday succeed him, especially because both his sons frequently squabbled about petty things. As they grew and their sibling rivalry escalated to the point where they were battling with weapons in the open fields, the emperor separated the two boys to avoid serious conflict. He sent Ebo to the east to take charge of the Shang, those who worship the Morning Star, or Antares, and Shichen to the west to take charge of the Shen, those who pay tribute to the Evening Star, or the Belt of Orion.

That these stars never cross orbits guaranteed that the quarreling brothers would remain forever separated.

When Ebo arrived at his assigned post, he noticed that the Shang lived without fire. He tried to steal fire from the star in heaven, but because its light was burning hot and moving constantly, he was unable to capture a flame and keep it burning. Then he had an idea. Why not bring a blade of dry grass along, catch it on fire, and keep the sparks lit until he got it back to earth? Then he could reignite the sparks into a flame. He succeeded—and they say that since then, people have been able to eat cooked food and find their way about at night by carrying lighted torches.

Nothing is written of the adventures of Shichen in the west, but we do know that neither brother acceded to the throne—that honor fell to their younger brother, Zhi. A Tang dynasty poem from the eighth century metaphorically describes the predicament of Shichen and Ebo with the phrase "We've lived our lives and haven't seen each other," like the never-meeting stars. And rulers of the Zhou dynasty (1046–256 BCE) used the story of the quarreling brothers to teach a moral lesson: the Shang dynasty that preceded theirs failed because its family members continually antagonized one another.

In the Americas, Orion also appears as a male figure, except that his name is Epietembo and he's missing one leg. To the contemporary Carib people of northern South America and the Antilles, Epietembo is the new husband of a young woman, Wawaiya, who is tempted by Maipuriyuman, a secret lover who becomes so smitten that he takes the form of a tapir just to be near her. (Tapirs look like a cross between a pig and a horse, and are known for foraging along sinuous muddy jungle trails—as well as for having long penises.) Maipuriyuman promises to assume human form if Wawaiya will follow him to the eastern horizon and ascend with him into the sky. As she collects wood for the fire, she secretly considers the proposal while Epietembo, none the wiser, gathers ripe avocados nearby. When Epietembo descends from one of the avocado trees, the young woman, by now determined to flee with Maipuriyu-man, grabs her ax, which has conveniently been magically endowed with extraordinary sharpness by her tapir lover, cuts off Epietembo's leg, and takes flight through the thick jungle with her new beau. But Epietembo recovers and, supported by a hastily fashioned crutch, seeks the whereabouts of his fleeing spouse, casting a trail of seeds from the avocados that sustain him in his pursuit.

After a winding journey along a maze of trails, Epie-tembo suddenly comes upon the couple making love. In a rage, he cuts off the head of the tapir-man and implores Wawaiya to return home with him. Instead she elects to pursue her lover's spirit into the heavens. Undaunted, the husband follows in hot pursuit. You can still see the three of them racing across the sky: Wawaiya is what we call the Pleiades, located next to Maipuriyuman's severed head, the Hyades, while Epietembo's bright red eye is the star Aldebaran, and the jilted husband's remaining lower limb is marked by bright blue Rigel, following close behind.

The eternal love triangle, a marriage broken via seduction: there is plenty to discuss when the morally fraught tale of Epietembo is told around the campfire, with the action animated in the winter-sky triad known to us as Orion, the Pleiades, and the Hyades. Like the Greek myth of Orion, there is a seasonal arc to the story. The encounter with the tapir-lover happens in the dry season, when avocados begin to ripen and wood is collected; the husband's frantic search occurs during the rainy season, when trails, more easily navigated by tapirs than people, run cold; and in the final act, which appears in the sky during the planting season, Epietembo finds the amorous duo, but only after the seeds of the avocado fall to the ground. The Pleiades rise before dawn in mid-June, followed by the Hyades, then by Orion. In this way, the seasonal rhythm of the constellations portrays a Caribbean fertility myth.

When the Lakota of the U.S. Upper Midwest cast their eyes on the Orion region, instead of the figure of a man, they see a hand. Orion's belt makes up the wrist, and his sword outlines the thumb. Rigel is the tip of his index finger, and the star Beta, borrowed from the constellation of Eridanus to the west, serves as the end of the pinkie finger. The configuration of the Lakota Hand constellation may be different from the Greek star pattern familiar to us, but the lesson about "The Chief Who Lost His Arm" is similar. As in the Chinese myth, it's all about how *not* to behave.

In the Lakota story, Fallen Star, an aspiring young warrior born of an earthly mother and stellar father, proposes marriage to the daughter of the head of a neighboring tribe. She agrees, but there are conditions. Earlier, her father had his hand snatched by the spirits of the Thunder People when he had become selfish and refused to share any of his harvest crop in a divine offering to the rain gods. The young woman will marry Fallen Star only if he can retrieve her father's hand. Taking up the challenge, Fallen Star travels from place to place among the Black Hills of what is now South Dakota, acquiring special powers from friendly spirits to assist in his planned getaway from the Thunder People, who, if they captured him, could bring violent storms and floods to the land. He collects a sinew; a live coal; feathers from an eagle, a swallow, and a wren; and most important of all, magical words and incantations that

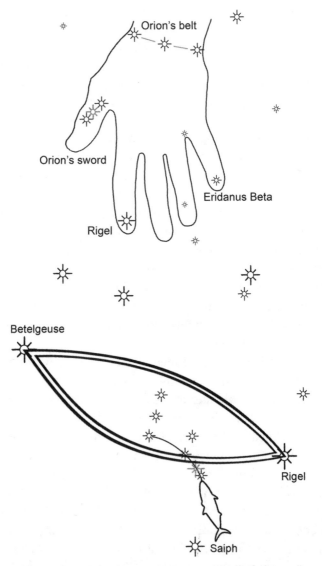

Four versions of Orion: the Lakota Hand, after R. Goodman, "On the Necessity of Sacrifice in Lakota Stellar Theology as Seen in 'The Hand' Constellation and the Story of the Chief Who Lost His Arm"; the Three Fishermen of indigenous Australia, after R. Norris

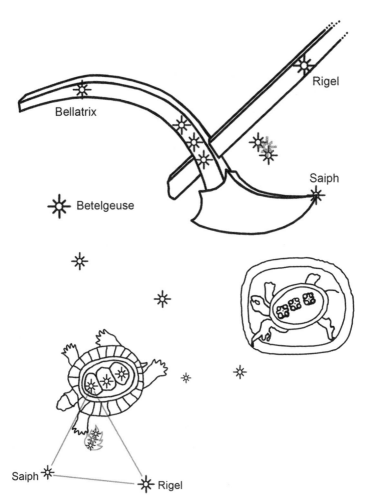

Rigel

Bellatrix

Saiph

Betelgeuse

Saiph

Rigel

and D. Hamacher, "Djulpan: The Celestial Canoe"; the Farmer's Plow of Indonesia, after G. Ammarell and A. Lowenhaupt Tsing, "Cultural Production of Skylore in Indonesia," 2210; and the Three-Stone Maya Hearth of Creation, with the Orion Nebula as the fire in the hearth, and the Belt stars comprising the back of a sacred turtle, after D. Freidel, L. Schele, and J. Parker, *Maya Cosmos*, fig. 2.14. (Redrawn by Julia Meyerson) will allow him to transform himself and so facilitate his escape. He also visits Star World, home of his ancestor star people, who offer him an escape shelter if he promises that his future son will someday pay them a visit as well.

Once Fallen Star reaches the land of the Thunder People, he manages to infiltrate them. Outwitting them with magic words that temporarily quell their thunder and lightning, he snatches the hand, which he promptly returns to the grateful chief, who vows never again to deny the gods their due. As promised, the young warrior is allowed to marry the chief's daughter, and she gives birth to a son a year later. That son will become the next chief.

Thanks to help from the younger generation, harmony between the Lakota people and their gods is restored. Transgressions are forgiven and there's a happy ending with a simple moral for a young audience: don't be greedy. But for more mature and observant viewer-listeners, there's more to the story of the lost arm than a lesson about the consequences of selfish behavior—there is also the seasonal timing of the Hand constellation in the sky. For the star pattern disappears in late winter, at the end of nature's reproductive cycle, a divine signal of the landscape's loss of fertility. The Hand then reappears in autumn, a sign that the sacrifices offered during the annual summer solstice ceremony, which reenacts the myth of the Chief's Hand, have been effective. Because the Lakota people have taken action, the renewal of life is assured in the next seasonal cycle. The old chief represents the previous cycle, the old year, while his grandson successor symbolizes the new year—which is ushered in by the generative power inherent in the human hand that casts seed over the earth. The daughter symbolizes the fertile earth mother, while the son she ultimately bears signifies, through the seed implanted in her, the regenerated life forms that emerge during the new year.

Indigenous Australians living in the outback tell their own story about bad ethical behavior involving parts of Orion. They call their constellation Djulpan, the Canoe. Once upon a time there were three brothers who went out fishing. They caught a kingfish and ate it, even though they knew it was against the law. For their violation Walu, the sun woman, sent a giant waterspout that carried the brothers and their canoe up into the sky as a reminder that during the fishing season everyone must follow the rules. You don't need much of an imagination to see the three brothers. Betelgeuse is the bow of the canoe and Rigel, the stern. Seated at the widest position are the stars of Orion's Belt, the three fishermen. If you look closely you can even see the forbidden fish trailing along on the line just below the Sword of Orion. We can easily spin this Orion tale to resonate with our contemporary moral code about not consuming endangered species such as sea bass. But this isn't what the Australians had in mind. The kingfish is not endangered. Rather, the brothers, being members of the Kingfish Clan, are strictly forbidden to eat the fish that is sacred to them. The patterns in the sky respond to our moral behavior on earth, but social rules regarding good behavior often differ from one culture to another. In the Aboriginal code of ethics, each animal family has its own human protector kin.

The contemporary Maya of Yucatán connect Alnitak, the belt star on the east, with Rigel and Saiph to form the three "Hearthstones of Creation." The fire and smoke in the hearth are what we recognize as the Great Nebula in Orion, visible to the naked eye as a faint fuzzy patch in the middle of the stellar triangle. The Maya story has deep roots. An inscription carved

on Stela C at the ninth-century BCE ruins of Quiriguá, in Gua-
temala, tells how the world was renewed by a trio of gods who
"planted the stones" after a previous race was destroyed by
flood:

> Three stones were set. They planted [the first stone],
> Jaguar Paddler, Stingray Paddler . . . He planted the
> [second] stone, First Black Chak (?) . . . The [third]
> stone was set [by] Great Itzamna . . . It happened at
> Lying Down Sky.

"Lying Down Sky" is the ancient Maya name for Quiriguá. The
three gods are those who paddle the canoe that carries the
dead over the river of the underworld, which is accessed via
the Milky Way.

The belt stars above the three hearthstones form the Maya
constellation of the turtle, the first form of life to emerge out of
the earth, before the sky was raised up. He carries the three-
stone hieroglyph on his back. There may be a metaphorical
connection between Orion's Belt and the seasonal maize cycle.
The belt would disappear during the first maize planting of
the Early Classic Maya Period (500–200 BCE), that is, from
late April to early June, and would reemerge when the maize
sprouted. When the crop was ready to harvest, in late September, the three stars would be aligned vertically above the horizon at midnight, resembling a fully developed maize plant. The
idea of three hearthstones continues to be important in Maya
culture. As a reminder of the sacred fire that nurtured the new
race of people (those of the present creation), a triangle of large,
round hearthstones marks the center—the cooking area—of
many contemporary Maya households.

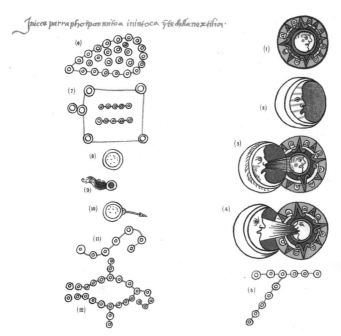

Aztec constellations from a Mexican manuscript. They include the
Pleiades (*top left*) and the sticks for drilling New Fire, consisting
of Orion's Belt and Sword (*bottom right*). The exact position and
number of stars in Orion's Belt, and of the star pattern surrounding
the Pleiades, may look unusual to Western stargazers; some cultures
don't rely on modern, scientific mapping techniques.
(Real Biblioteca [Madrid], II/3280, 282r–282v)

     The Aztecs of Mexico imagined a constellation they called
Mamalhuaztli, the "Fire Drill." They identified it with the Belt
and Sword of Orion. Mamalhuaztli are the sticks used by the
Aztec ruler to rekindle the fire that was lit according to a cal-
endar cycle of fifty-two years, known as the "Binding of the
Years." This cycle, a hefty human lifespan back in the day, is

timed to coincide with both the sacred count of 260 days and
the 365-day seasonal year. A Spanish chronicler of the Aztecs,
Bernardino de Sahagún, tells us that when the Pleiades star
group was positioned overhead and the sun lay directly under-
foot, "it was a sign to the anxious waiting multitude that the
world would not be destroyed and that a new [cycle of life]
would be granted to mankind." Aztec skywatchers recognized
that when this event took place, the fire sticks would be located at their highest point in the sky. What better time to
beckon new light to return to the world than when the sun is
farthest from us? On this occasion everyone would toss away
their sleep mats, break their dishes, and destroy all other household implements used during the previous cycle. They would
then assemble at the Great Temple of Tenochtitlan, the Aztec
capital (today Mexico City), where a "new fire" was drilled. In
the spirit of renewal of life, each householder would immerse
a pine stick into the communal fire and carry the flame home
to rekindle their hearth. They would then craft new ceramic
bowls and plates, sew replacement mats, and, like those of us
who make New Year's resolutions, start their lives anew.

Though they emanate from widely disparate cultures,
many of the stories told in this chapter have elements in common. The Greek Orion, One-Legged Man, Quarreling Brothers,
and Australian Fishermen stories all encourage conversations
about exceeding or breaking ethical codes. The Fallen Star, Fire
Drill, and One-Legged Man myths, too, deal with the idea of
renewal and the domestication of time. But I find a delightful
surprise in the story of the great cosmic fire of Maya creation.
Modern astronomy teaches us that the Orion Nebula is actually a vast cloud of dust and incandescent hydrogen gas, a ten-
thousand-degree fire where stars are still being born, many
of them surrounded by cosmic disks out of which future life-

bearing planets are likely to emerge. This star story, then, imagined so long ago and retold for generations, holds meaning for truth-seekers even today, animating discussions and inquiries that will continue long into the future.

2

One Pleiades Fits All

To the Navajo, Black God is the equivalent of the Greek Prometheus, bearer of fire to the world. He is said to have been one of the deities who inhabited the first of four worlds from which the ancestors emerged. Black God wears a mask darkened by charcoal and marked with several white symbols. A long line bisects the face of the mask. At its base lies a mouth, a round sun disk. A crescent moon marks the top of the nose, and over the left eye we see the unmistakable image of the glittering little cluster of stars we call the Pleiades, or Seven Sisters.

A legend told by the Navajo to Father Berard Haile, who served at Saint Michael's Franciscan mission in northern Arizona in the early twentieth century, tells of Black God entering the *hogan* (house) of creation. When the other gods pointed out that several crystals were attached to his ankle, Black God vigorously stomped his foot, causing the crystals to fly up to his knee. He stomped his foot again and the crystals rose to his hip; after a third time, the crystals landed on his shoulder. When he stomped his foot a fourth time, the little crystals that comprise the Pleiades lodged next to his left temple. "There," he said, "they shall stay!"

This impressive feat convinced the ancestors that Black God possessed the power to beautify the dark sky. They encouraged him to continue making star patterns out of crystals and placing them in the heavens. He opened his deerskin pouch and withdrew handfuls of crystals and carefully decorated the sky with the "Male and Female Who Revolve" in the north (the Big Dipper and Cassiopeia), "Rabbit Tracks" and "Man with Legs Ajar" (Scorpius and Corvus) in the south, and all the other constellations. Then he set them aglow by supplying the "Igniter Stars," Dilyehe (the Pleiades), with his eternal flame. They say that Black God's meticulous placement of each constellation gave us a kind of *sa'a naghai bik'e hozho,* which according to one Navajo elder is "the holistic and ordered essence of life that encompasses the universe . . . the life force which is the reason for being and becoming."

The map visible on the face of Black God's mask hints at another reason for the Pleiades' nearly universal recognition: their proximity to the ecliptic—the path followed by the sun, moon, and planets among the stars. Like a bustling diner alongside a busy highway, the Pleiades play host to a vital stream of cosmic traffic. The crescent moon on Black God's masked forehead is situated exactly the way it appears when the sun is positioned below it, with the cusps of the crescent pointing away from the source of light setting in the west. (If you imagine that the lunar crescent is a bow, then the arrow it discharges aims at the sun.) The line that connects them is the ecliptic, which can also be understood as the plane of the earth's orbit, traced out by the constellations of the zodiac. If we think of Black God's face as a sky clock in addition to a map, the first

The Pleiades star cluster, which adorns Black God's face.
(NASA, ESA and AURA/Caltech)

crescent moon low in the west signals the month when the Pleiades make their final appearance in the sky at sunset—the last
month of winter that anticipates the onset of greater daylight.

"I cannot explain its significance, and why of all stellar
objects this minute cluster of stars of a low magnitude is more
important than other stellar groups is not clear to me." So wrote
Jesse Fewkes, indefatigable archaeologist-explorer of the southwestern United States in the 1890s, describing in his field notes
an invocation made during the Navajo Nightway Ceremony
to the gods who represent the four quarters of the world. The
ceremony was held on the first evening every winter when the
Pleiades would appear in the east after sundown.

Unlike Orion, whose parts make up different constellations in different cultures, the Pleiades, though often confused with the Little Dipper, are recognized universally as a little cluster of six, seven, or eight stars. Each individual member star may not be so bright, but their combined light, spread over an area of the sky about the size of the full moon, catches the eye of anyone who casually looks skyward.

The Pleiades are mentioned three times in the Bible. Muhammad and Plato wrote about them. So did Milton, Byron, Keats, and Tennyson. "An image of Elysium lies: / Seven Pleiades entranced in heaven / From in the deep another seven," remarked Edgar Allan Poe in his poem "Serenade." Amy Lowell created an entire poem about them, "The Pleiades." Their image is prominent on the 3,500-year-old Nebra Sky Disk, the world's oldest star map, and you can spot the same design on the grill of every Subaru vehicle. Each star is said to represent one of the companies that merged to form the corporation. (Subaru, which means "united," is the Japanese word for the Pleiades.)

When Carib Indians saw these seven stars, they imagined human entrails. Their Pleiades story is reminiscent of the biblical Cain and Abel. The ghost of Tumong, who was murdered by his brother, who desired his wife, haunts his slayer until he is reburied and his guts are scattered in the sky as a reminder of the most evil crime of fratricide. In medieval Turkey, the Pleiades were a military formation ready for ambush. Norse mythology portrays them as a hen with her chicks. In places as far removed as the Andes and Ukraine they are a storehouse, which fits with how each year at around harvest time they reappear in the east after the sun goes down.

Inca descendants in the Peruvian highlands also time their most important annual festival, Q'olloy Rit'i, or Quyllurit'i ("bright white snow"), to the annual disappearance of the Pleiades. On the day of the first full moon following their reappearance after a forty-day absence in the sun's glare, tens of thousands of pilgrims undertake a journey to the high peaks. There they kneel to receive the first gleam of the rising sun that brings with it a return to order and the start of the new year.

The Iroquois, People of the Longhouse, tell a sad Pleiades story about child abuse and what happens when parents lose respect for tradition. Having traveled for many generations, the ancestors finally settled in a new land with lush forests and many clear lakes. There they created spaces to plant their staple crops of corn, beans, and squash. In their fields they constructed a longhouse to shelter their growing extended families. But as they prospered, they began to lose sight of the ancient rituals of Thanksgiving taught them by their creator. They began to quarrel. Some even left their homes. They taught their children not to speak or play with anyone who wasn't part of the immediate family. Children who disobeyed were whipped and denied food. Older siblings recalled stories, told by their grandparents, of better times when great feasts had been held and everyone had sung, danced, and feasted in honor of Mother Earth. During that earlier era, parents had truly cherished their children as gifts from the creator.

One day a group of seven children decided to sneak away into the woods and stage their own Thanksgiving ceremony. They sang the old songs and danced the old dances they remembered. They shared bits of food brought by each participant. But those who were caught sneaking away at night were punished harshly. Some were even tied to their beds after being whipped and denied supper. Still, the children kept their promise to meet in secrecy.

During one storytelling session at a forbidden ceremony, one of the children related a tale told by her grandmother about a special place called Sky World. It was the original home of the ancestors of the Longhouse People, and if the Iroquois children went there, they would be welcomed home by the First People's ancestors. One night, while the seven children were dancing and singing around the fire, they asked Heaven Holder to take them up to Sky World. They did a special dance and sang a special song. Just as they began to feel themselves being lifted into the sky, a group of parents who had followed their trail into the woods showed up. They first cried out angrily to the children for disobeying them, but their anger turned to despair as they stood in awe watching their young ones rise slowly into the heavens. When the parents heard the songs the children were singing, they suddenly realized how badly they had harmed their own flesh and blood. They wept and called out, begging the children to come back to this world. But higher and higher the seven children rose. One of them, whose parents had never actually struck him, turned his head to look downward; there he saw his mother pleading with him far below, her arms raised up. The boy stopped singing and began to fall; he plummeted faster and faster until he became a luminous streak. The other dancers left their families and permanently disappeared into the heavens.

After this, when parents witnessed a shooting star they would vow never to strike their children again. And when the People of the Longhouse go outside in the early evening at the time of the harvest—Thanksgiving—they look to the northeast for the Pleiades, that little cluster of stars in Sky World to which the young ones had ascended, then tell their children how much they love them.

A story told to young women from the bush of Western Australia centers on the KungaKungaranga, the Seven Sisters. "They are your relatives," says a girl's grandmother; "they come

from the same country as you and me." When the Aboriginal
people speak of their country they mean the land, the sky, and
the emotional connection with everything around them. Long
ago in Dreamtime, the Seven Sisters would come down from
the sky for a visit. They always landed in the same spot on the
hill where you can still watch them set. There's a cave on that
spot, a secret passage the Pleiades would enter and use as a
temporary home when they stayed a while here on earth.
During one visit, the KungaKungaranga went out hunting and
gathering for dinner. An old man, desperately in search of a
wife, followed them unnoticed. He stalked the sisters all the
way to the side of a creek where they set up a temporary camp.
Suddenly the man jumped out from behind a bush and as the
girls scattered, he grabbed the youngest one. Her frightened
sisters fled to their hilltop and quickly ascended into the sky.
The youngest sister tried to fight off the old man's grip and,
managing to break away, ran toward the hill shouting out to
her sisters, not realizing they had already departed. As she ascended, the Old Man followed her up into the sky. You can still
see him chasing her. She is the faintest KungaKungaranga and
he is the Evening or Morning Star, ever in hot pursuit. There he
goes, say the elders, still chasing the Seven Sisters.

Many are most familiar with the Pleiades as the seven
daughters of Atlas, a titan condemned to uphold the sky, and
the sea nymph, Pleione. The beautiful daughters were pursued
by the ardent womanizer Orion. For their protection, Zeus felt
impelled to change them into doves and then, after Atlas asked
for a more secure disguise, into stars. You don't have to look
very far to the east to notice that Orion, like the old man in the
Australian myth, is still in eager pursuit.

Sky stories remind us of the world we came from and the world we now inhabit. They tell us when we need to recalibrate our moral compass, and they warn of real-life dangers that confront us. But they also serve the practical function of displaying the passage of time, which in turn has helped humans create a seasonal calendar. Hesiod's *Works and Days,* written in 700 BCE, is a Greek poem intended to be recited to an audience on formal occasions. Hesiod was a hard-bitten farmer who labored all his life scratching out a living on the rockbound turf of the Peloponnese region. As he aged and grew weary, he realized that his wayward younger brother, whom he'd hoped to groom to take over the farm, needed specific instructions on the seasonal schedule of activities. And so, it is said, Hesiod wrote his poem.

The Pleiades play a more important role in Hesiod's verses than anything else in the sky; he mentions them at least five times. For example, here's how Hesiod describes the period between when the Pleiades first appear above the eastern horizon following sunset and when they drop out of view in the west before sunrise, and how these timings should guide the work of harvesting and plowing:

> At the time when the Pleiades, the daughters
>   of Atlas, are rising,
> begin your harvest, and plow again when they
>   are setting.
> . . . .
> The Pleiades are hidden for forty nights and forty days,
>   and then, as the turn of the year reaches
>   that point
> they show again, at the time you first sharpen
>   your iron.

Hesiod also uses the celestial departure of the Pleiades to fore-
cast stormy weather:

> But if the desire for stormy seagoing
>   seizes upon you:
> why, when the Pleiades, running to escape
>   from Orion's
> grim bulk, duck themselves under the misty face
>   of the water,
> at that time the blasts of the winds are blowing
>   from every direction,
> then is no time to keep your ships
>   on the wine-blue water.

Today, farmers in the high Andes also look to the Pleiades to predict the weather. They say that if the Pleiades are
clear and bright in the predawn sky, good weather is in store
for the coming months. But if they are dim, expect a meager
potato harvest: it will be better to delay planting because pre-
cipitation will be late and sparse. Modern scientific studies ver-
ify that this technique, in indigenous use for over four hundred years, works quite well as a predictor of El Niño, which
accompanies droughts in the Andean highlands. Atmospheric
scientists note that the primary visual attributes of the Pleiades
used by natives—overall brightness and size, date of first visi-
bility, and appearance of the brightest member—all correlate
with the visual occurrence of the high transparent clouds that
appear months before an El Niño year. Folkloric practice meets
scientific knowledge!

The Pleiades agri-clock also ticks for contemporary Indo-
nesian rice farmers, who are well aware that successful crops
can be anticipated by tuning in to the stars' dance across the

sky. The Pleiades form a pattern the farmers call Bintang Weluku (plow stars). They mark the new agricultural year by their first appearance in the east in the early morning sky, which in Indo-Malay latitudes occurs in late December, just after the winter solstice. This signals farmers to initiate the month of plowing for rice planting. The sky acts out the scene as well. Imagine that Orion is a plow: the handle is Rigel, the base is Bellatrix, and the belt and Saiph make up the plowshare, or tip of the shoe (see the drawing in Chapter 1). The sky image of the plow stands upright over the east horizon, ready to overturn the soil beneath it, as the Pleiades appear, announcing the work that is about to commence.

The same seasonal event, the early morning rise of the Pleiades (*isiLimela,* or "digging stars"), marks the time the South African Zulu start digging or hoeing. Farmers spend a lot of time arguing over exactly when a first sighting occurs. Some look for the first Pleiad, others insist on seeing the whole group. Some claim to view ten or twelve member stars. (I have seen a dozen in the clear air of the high Andes.) Zulu farmers often set up contests to see who possesses the greatest visual acuity and the most astronomical knowledge. The neighboring Xhosa tie the plowing or the "coming out" of the earth with that of young men, who are circumcised and ceremonially celebrated as full adults during the same month. One Xhosa man gave his age in manhood years by how many first winter appearances of the Pleiades had elapsed since his coming-out ceremony.

The Pleiades also play a supporting role in an Aztec story of intrigue I call "Murder in the Sky." This myth, propagated by imperial rulers during a period of conquest, features the god of sun and war, Huitzilopochtli, who was the patron deity of their capital, Tenochtitlan. Aztec warriors killed in battle

followed Huitzilopochtli across the sky so they could be transformed into hummingbirds in the hereafter. Legend has it that
Huitzilopochtli's mother, Coatlicue, the serpent-skirted earth
goddess, came upon a colorful bundle of hummingbird feathers. She stuffed them into her bosom for safekeeping, and soon
found herself mysteriously pregnant. This upset Coatlicue's
daughter, the moon goddess Coyolxauhqui, who became jealous of her unborn sibling and plotted to kill her mother.
Coyolxauhqui persuaded her four hundred brothers, the Pleiades, to join her in the scheme, but an informant among
them warned the unborn child, Huitzilopochtli, who suddenly
sprang from his mother's womb fully grown and armed for
battle. He chased his sister down, butchered her, and hurled
her head into the sky. Then he pursued the four hundred
brothers and scattered them.

Excavated in the 1980s, the Temple of Huitzilopochtli is
the tallest building in ancient Mexico City. A carved stone
on its summit records Huitzilopochtli's mythical date of birth.
More recently the Stone of Coyolxauhqui, a huge disk displaying the dismembered body of the moon deity, was excavated
at the base of the temple. It was carefully positioned precisely
along the axis where the sun rises over the top of the temple
on the first day of spring, when the Aztecs conducted rites to
the sun-war god. The lithic representation of Coyolxauhqui's
corpse was placed at the bottom of Huitzilopochtli's temple not
only because he was the one who slew her, but also because this
is the place where the bodies of sacrificed captive warriors
were divided among their conquerors. The message of the imperial state is made resoundingly clear in the star story: the
Aztecs would do to those whom they conquered what Huitzilopochtli had done to his sister, Coyolxauhqui.

The dismemberment rite has a celestial parallel, where

the sun's victory over the moon and the Pleiades is periodically reenacted for all Aztec citizens to witness. Unlike Huitzilopochtli, the sun, who passes into the underworld in the west and reemerges anew in the east the next day, the moon dies and becomes dismembered when the sun catches up with her in her early morning waning phases, only to be reborn and replay the chase next month. He continues on his annual course and later "scatters the four hundred Brothers," to borrow the chronicler Sahagún's words. Just as the birthdate representing the solar deity needed to be made visible above the plaza in front of the temple, where the actual rites to Huitzilopochtli were conducted, the mythic axis between the male and female deities, who confront one another, needed to be symmetrically aligned east-west with the cosmos. He (the winner) is like the sun; she (the loser) is like the moon. He lies at the top of the temple, she at the bottom; Huitzilopochtli in the east, Coyolxauhqui in the west.

Whether it be a tale simply told, like the Iroquois or Aboriginal Australian stories about the need to protect their young, or a grand opera like "Murder in the Sky," the effectiveness of any performance depends critically on set design—the arrangement of elements on the stage in ways that enhance the actors' speaking or singing roles. Unlike those of us who worship in temples, mosques, and churches, the Aztecs, who lived in a tropical climate, conducted their holy rites out of doors, under an open sky that took the place of story-enhancing elements we are more familiar with, such as stained glass windows and painted altars. In a city of more than a hundred thousand, the Temple of Huitzilopochtli and its environs once served as the grandest of all stages for the reenactment of Aztec military history. It was there, in that urban ceremonial center, that messages and meanings were conveyed to the believers by the set decorators—called priests, urban planners, architects, or something else—who astutely saw to it that sacred objects, both stars and stone carvings, were brought together in the right place at the right time.

3

Zodiacs Around the World

To celebrate his birthday, the Chinese Jade Emperor decided to create a zodiac. For the twelve available slots he auditioned a host of animals: the winners would be the first dozen to cross a river. The cat and the rat, once good friends, and both terrible swimmers, decided to hitch a ride on the ox's back; but halfway across the river the selfish rat pushed the cat into the water, eliminating him from the competition and making an enemy forever. Just as the ox approached the finish line on the opposite side, the rat jumped off and beat him out for first place. The ox, never an animal to complain, came in second, a nose ahead of the tiger, who, though speedy on land, had become horribly weighed down in the river by his wet fur. The rabbit was a favored winner because of his natural swiftness, but he was delayed because he decided to make the crossing by hopping from rock to rock to avoid getting wet. Near the far shore he slipped and fell in, but he was lucky enough to grab hold of a floating log that finally brought him safely to shore. By taking such a circuitous route, the rabbit finished a disappointing fourth.

Much to the surprise of the emperor, the dragon came in

only fifth. When asked by the Jade Emperor why an animal with such magical powers could not have done better, the dragon replied that he had felt compelled to stop to save a group of farmers trapped in their burning field. (He blew out the flames with his powerful breath.) The Jade Emperor complimented the good dragon for making a wise choice. It looked as if the horse was headed for sixth place, and the sneaky but clever snake, which had wrapped itself around the horse's foot to hitch a ride, was destined for seventh. But careful observers noticed that the snake had jutted out his head just enough to touch dry land ahead of the horse, so their finishing positions would be reversed.

With five slots remaining in the competition, excited spectators cheered the monkey, the rooster, and the sheep as they paddled furiously, neck and neck, toward shore. Suddenly at the last moment, the monkey and rooster decided to concede eighth place to the sheep because they felt she had bonded most harmoniously with all the competing animals. The soaking wet dog came next, out of breath with his tongue hanging out. Just like a dog, he'd spent too much time playing in the river. And twelfth place? No animal was in sight, and the Jade Emperor thought maybe it was time to call it quits. But then the crowd heard a distant oinking sound. It was the lazy pig who had stopped to eat, then indulge in a postprandial wallow in the mud, before finishing in time to take the last place in the zodiac.

In the Great Race tale it's easy to see how the Chinese storyteller comments on the moral behavior of human beings with a cosmic backdrop and with animals as stand-ins—clever snake, playful dog, lazy pig, and so on. It's also fitting that the Chinese arrangement (Sheng Xiao) is a parade of animals; the word *zodiac* comes from Greek for "circle of animals." Astronomers define it as an 18-degree-wide band that passes all the way around the sky, divided into twelve 30-degree segments. The Western zodiac is a star-studded route that circles the sky, marking out the path the sun traverses in a year (the ecliptic), and the arc of the moon in a month. It also charts the course of five other bright lights that move back and forth in the night sky—the planets Mercury, Venus, Mars, Jupiter, and Saturn.

Though modern astronomers know that the earth orbits the sun each year, we actually *see* the sun pass from west to east among the stars during that same period. (We may witness the sun move from east to west, together with everything else in the sky, in a day, but in modern terms we attribute that motion to the twenty-four-hour rotation of the earth.) Similarly, we think abstractly of the moon orbiting the earth in a month (close to the plane of the earth's orbit), though we witness it moving among the constellations, also from west to east in the same period. Likewise we know that the planets orbit the sun, more or less in the same plane, each in its own period, even though we see them taking the same general west to east course along the stellar highway, except for when they go backward (retrograde) for brief periods. This happens when the earth passes them or they pass the earth in orbit. You get the same effect when you pass a vehicle traveling in the same direction along the highway. Relative to the distant background of trees and hills, it momentarily seems to be moving backward.

Western tradition has it that the sun, the moon, and the wanderers—as the word *planet* implies in Greek—traveled "the way of Anu," the ancient Sumerian name for the zodiac. Bright stars lit the way, and their positions were interpreted by the counselor gods or consultants, who were in charge of surveying what happened in their designated zones of the world and advised the gods about future undertakings. The northern region of the zodiac, near the Tropic of Cancer, constituted the way of Enlil or Bel, lord of the earth, while the southern sector, the Tropic of Capricorn region, was the way of Ea, god of the waters. Anu, the sky god, ancestor of all deities, regularly took to the zodiacal skyway to inspect its stations and consort with the planetary gods who visited them.

Taurus, the bull, whom Orion seems to be attacking (for no apparent mythological reason), was one of the earliest members of the Greek celestial zoo. Along with Leo, the lion, and Scorpius, the scorpion, the bull's effigy appears as a heraldic figure on boundary markers and cylinder seal impressions dated back to 3200 BCE. His placement in the heavily trafficked section of the sky where the Milky Way crosses the zodiac was bequeathed to him by Zeus in return for carrying Europa, the moon goddess, over the sea from Phoenicia to her home in Crete. A livelier version of the story has Zeus falling in love with Europa. To make an impression, he approaches her in the form of a white bull. She is so taken with the animal's beauty that she climbs on its back and gets carried away to the Aegean island.

Leo was king of all the beasts, until Heracles slew him with his bare hands; and we've already met Scorpius, destined to reside at Orion's antipode. Virgo, the virgin, is one of three humanoid constellations. Given her proximity to Libra, the scales, she represents Dike, or justice, another of Zeus's daughters. They say that Virgo once lived among us here on earth, but she became so disgusted with the way the human race had declined from the Golden Age of the past that she gave up trying to uphold the law. So she exiled herself, first to the distant mountains and then, when mundane affairs got even worse, all the way to heaven. Aquarius, the water bearer, looks like a man pouring water out of a jar. He is one of three consecutive "watery" constellations in the zodiacal lineup. The other two are the half-fish, half-goat Capricornus and the twin fish Pisces. The three hang together consecutively in the sky as a reminder that the sun passes through their houses during the rainy season, a vulnerable period that ends when the sun leaves them and reappears over the eastern horizon just before dawn. Some of the Hebraic names for months, such as Nisan (sacrifice) and Iyar (blossom), still reflect traces of activities that took place in the local civic, ritual, and agricultural calendar.

Aside from the motivation to enshrine traditional stories with lasting meaning, a culture's desire to construct a zodiac seems to be enhanced by the natural human tendency to lend worldly attributes to the forces of nature. Even today, astronomers speak of stars being born, black holes devouring stars, and supernovae undergoing one last cataclysmic fling before dying. It's not surprising, then, that pastoral people all over the world who watched for the signs of the seasons—where the sun resides during each of the twelve lunar months over the course of the year—would describe the sun's place on the celestial roadway with names that evoke natural occurrences.

In a word-association test, the standard response to "zodiac" is usually "astrology." Whether or not you choose to believe your horoscope or love pairings, astrology is still part of a shared human desire to find order in the world, to search out the underlying harmony that we believe must exist. It is our nature as human beings to ask questions: What's the source of authority and control—the power of one individual over another, of rulers over their subjects, of people over nature, of nature over people? How can we know what lies ahead? Of all the natural media for conveying pristine order, perfection, and certainty, none exceeds the capacity of the sky. Nothing is more predictable than the paths that the sun, moon, and planets follow along the zodiacal roadway. You can depend on the signs in the sky. That's why we've always turned to it.

But attempting to acquire foreknowledge by witnessing celestial phenomena is no mean task. It demands not only a high degree of familiarity with the ways of the transcendent, but also great skill and persistence by the professional skywatcher. To judge from an inscription on a statue of Harkhebi, a famous Egyptian astrologer, divining by the stars must have been an exalted profession, one to be undertaken by a "hereditary prince and count, sole companion, wise in sacred writings, who observes everything in heaven and earth, clear-eyed in observing the stars, among which there is no erring; who announces rising and setting at their times, with the gods who foretell the future."

Ancient writings also reveal that the life of an astrologer was far from stress-free. You can sense tension in this passage found in the diary of Minnabitu, an Assyrian court astrologer active in the seventh century BCE who is clearly very worried about job security:

> The king has given me the order: Watch and tell me whatever occurs! So I am reporting to the king whatever seems to be propitious and well-portending [and] beneficial for the king, my lord [to know] . . . Should the king ask, "Is there anything about that sign?" [I answer], "Since it [the planet Mars] has set, there is nothing . . ." Should the lord of kings say, "Why [did] the first day of the month [pass without] your writing me either favorable or unfa-

vorable [omens]?" [I answer], "Scholarship cannot
be discussed [heard] in the market place!"

The bold astrologer further laments: "Would that the lord of
kings might summon me into his presence on a day of his
choosing so that I could tell my definite opinion to the king my
lord!" Poor Minnabitu!

   We sense tension as well in the words of a Babylonian
priest, his eyes turned skyward, who pleads poetically with the
gods and offers an animal sacrifice for good measure: "O Pleiades, Orion and the dragon, . . . Stand by, and then, . . . Put
truth for me."

   Whether the errant diviner apprehended the vital signs
of nature in time we can only guess. Perhaps he consulted the
list of Mars's positions in the zodiac but chose the wrong day
of the month. Maybe he confused the corresponding events on
earth written against them. Or was he simply too overworked
and exhausted to adequately perform his task? These poignant
quotations portray the ancient astrologer not as some of us
might be inclined to see him—as a charlatan—but instead as a
helpless spectator trying very hard to follow the complex rules
of his learned profession. At least in the eyes of his king, the
court astrologer scarcely seems an arrogant channeler of true
knowledge, a wielder of real power. In some cases, the word
from on high sounds clear enough, but in others, like "Put
truth for me," it's rather opaque. But then, what successful religion doesn't grapple with the unanticipated, the novel? Religion is not in the business of giving precise answers to all the
questions that matter to us.

   A heavy price was paid when the welfare of the state depended on the court astrologer's action and he failed to deliver.

Ho and Hi, a pair of second-millennium BCE Chinese astronomers, were said to have been executed after they messed up.
Centuries later, their fate became commemorated in a ribald
rhyme:

> Here lie the bodies of Ho and Hi
> Whose fate though sad was risible,
> Being hanged for they could not spy
> Th' eclipse which was invisible.

A popular version of the story has it that they were drinking
on the job. The safest scholarly opinion is that they acted in-
appropriately when the eclipse occurred.

As above, so below—the logic behind stellar divination
is really quite straightforward. Through everyday experience a
careful observer can easily become aware that the cycles of the
sun and the moon are correlated with the seasons, the tides,
the menstrual cycle, and other biorhythms. It seems to follow,
then, that if we watch the sky carefully enough, we may discover associations between the most precisely predictable occurrences on nature's stage, such as eclipses or appearances of
Venus as the Morning Star, and more unpredictable ones, like
a plague or the arrival of locusts. These sorts of questions were
very much on the minds of antiquity's courtly timekeepers,
the pre-scientific sky specialists who composed and dictated
the contents of astronomical tablets to their scribes. The language of the dialogue between mortal here below and transcendent there above consisted of offerings and incantations,
not scientific experiments; the implements of communication
were charm and amulet, rather than compass and telescope.

If zodiac and astrology go together, another entry in the
word-association test has to be *horoscope*. We owe that to the

Greeks. *Horoscopus* means "I observe the hour," or colloqui-
ally, "I watch what rises"—and it refers to the art of predicting
the general patterns that are supposedly preprogrammed to
occur in your life based on an examination of the celestial
bodies coming over the eastern horizon when and where you
were born. Unlike the Babylonians, whose astrology was concerned with what might befall an entire state, the Greeks, reared
in a democratic system, believed that everyone was entitled to
their own personal horoscope. Imagine the journeyman astrologer engaged in a consultation in the crowded Athenian
agora, or central public space, 2,500 years ago. The client might
ask, "Will my sister's pregnancy end in a healthy birth?" "Can
I expect better weather for my crops next month?" In the Greek
culture, everybody had a right to knowledge about the future.

The Chinese actually devised three zodiac-like arrangements. One comprised the animal signs described earlier, and
marked a twelve-year (rather than twelve-month) cycle, with
each animal representing a year. Another involved twenty-
eight signs, arranged along the equator rather than the ecliptic, to keep track of the motion of the moon against the starry
skyscape. And the third divided the heavens (also along the
equator) into four directions, each represented by an animal
talisman, or mascot: Azure Dragon of the East, Vermilion Bird
of the South, White Tiger of the West, and Black Turtle of the
North. Each region had its own assigned characteristics—a
season, a color, an element, and so on.

The Chinese lunar zodiac, or *xiu,* appears on star maps
from fifth-century BCE tombs. Its signs are named after zoo-
logic and anthropomorphic body parts, such as beak, stomach,
wing, heart, and gullet, or after domestic items, like harvester,
house, well, ox, and winnower. Still other moon houses, like
ghost and triaster (the three stars that make up the belt of

Tang dynasty mirror displaying all three Chinese zodiac-like
arrangements, from the inside outward: the four directions,
twelve year signs, and twenty-eight lunar mansions.
(Granger, www.granger.com)

Orion), are more abstract. Though the system is complicated,
these names turn out to be neither obscure nor irrelevant in
their real world. For example, the mane and yak tail relate to
events pertaining to warriors, the net to hunters, the lasso to
prisoners, the stomach to matters of the warehouse and gra-
nary. The turtle beak governs the harvesting of wild plants,

while the ghost is capable of detecting cabals and plots against the emperor.

With each marked mansion came a specific omen. Much as in Babylonian astrology, these statements pertained to the domain of the Chinese emperor, the Son of Heaven, and the influence of celestial forces. Thus, "When a wise prince occupies the throne the moon follows the right way"; "When the high officials let their interests prevail over the public interest, the moon goes astray toward north or south"; or "When the moon is slow, it is because the prince is being rash in meting out punishments."

The animal year sign was the essence of *taiyang,* the force that sustains life, benevolence, and virtue—all obvious qualities of the emperor. But it could also reveal his shortcomings. Any changes in the sun's appearance presaged a change in the state of the empire. In times of war, for example, a change in the sun's color could augur defeat in battle, but in times of peace it might signify the death of a noble. Traveling its own highway, the moon was the sun's counterpart—not its opposite, but its complement. Its essence was *taiyin,* a force usually associated with the empress and her qualities. It too merited careful observation, for under a wise empress it kept its course. It moved south or north when the rules of punishment were not correctly applied, and a sudden change of color could mean the empress had behaved foolishly.

Following the courses of the planets, especially Jupiter, was also important in Chinese astrology: planetary advances and regressions, as well as their rapid back and forth movement among the zodiacal constellations, seemed to portend events in the human realm. Specific words were used to describe one planet passing around another, just like people descending from above or moving upward from below; and it was considered essential to note when one planet rushed by another or concealed another, when two moved in opposite directions along the same line, or when they covered each other, joined then separated, or hit each other. Leaders of the Ming dynasty, for instance, viewed a great planetary massing in 1524 CE as a heavenly mandate foretelling the end of their rule. In that year Mars, Venus, Jupiter, and Saturn dodged back and forth, then coalesced in the Black Turtle's quadrant. Were the "emperor's minions," as the Ming astrologers called the planets, gathering in the supernatural realm to discuss political changes that they might enact in the human world below? Ming historical records tell of an earlier close planetary gathering involving the same planets more than twenty-five centuries earlier, in 1059 BCE in the Vermilion Bird quadrant. They timed this event with the Zhou overthrow of the earlier Shang dynasty. Chinese historians speculate that the Ming may have known as well of a similar event that had happened in 1579 BCE. All three fit into an ultra-long 516-year pattern of planetary portents of great change, equivalents of the Star of Bethlehem, which according to Christian beliefs signaled the coming of Jesus as Christ. Some astronomers identify that event with the close conjunction of the bright planets Jupiter and Saturn in the constellation of Pisces at the time of the Christian savior's birth.

The medieval courts of Europe were as well populated with astrologers as those of the Chinese palace. Bishop and king, priest and prince—any official who wanted to know what would take place in heaven, and precisely when it would happen, consulted these sages. Cecco d'Ascoli, astrologer of the fourteenth-century court of Florence, was prominently involved in anticipating planetary conjunctions. He was a member of the Franciscan Order and special adviser to Florentine medical doctors. "A doctor must of necessity know and take into account the nature of the stars and their conjunctions," he tells followers in his book *Astrological Principles.* He goes on to list all the plants and herbs associated with each planet so they might be administered at the proper time. Sadly, power and fame led d'Ascoli to exceed the limits of his discipline. When he began to dabble in off-limits astrological predictions based on the birth of Christ, the coming of the Antichrist, and the end of the world, he landed in front of the inquisitor. In 1327 the Catholic Church burned d'Ascoli at the stake, whether for astrological malpractice or because of political intrigue we can't say. One of his judges, the bishop of the city of Aversa (also a Franciscan), regarded d'Ascoli as an ally of the rival city of Cesena, which in turn supported the breakaway faction of Franciscans to which d'Ascoli belonged.

Those of us in the habit of seeking rational explanations for physical phenomena can't fathom how closely the medieval mind connected worldly affairs with celestial movements. In 1348 the University of Paris medical department reported that at one o'clock in the afternoon on March 20, 1345, a conjunction of Mars, Jupiter, and Saturn in Aquarius had portended one of the world's most monstrous occurrences. How would it happen? Jupiter, warm and humid by nature, would draw evil vapors out of the earth, while hot and dry Mars would ignite them, and evil Saturn would spread them over the human world. Over the next three years the Black Death killed nearly half the population of Europe. Though we may have traded in a plague induced by star-crossed planets for a deforested Amazon, a disappearing ozone layer, and elevated global temperatures, some are still left wondering whether modern science will end up having any more control over the destiny of our planet than medieval astrology.

In 1542, when Spanish conquistadors entered the Maya capital city of Mayapan in northwest Yucatán, they were closely followed by the Roman Catholic priests, whose job was to convert the infidels to Christianity. They began by demolishing native places of worship and burning any "codices"—lime-coated, bark-paper books used to schedule astronomically timed periods of worship—that they could find. One missionary boasts of a huge bonfire, fueled by piles of these folded screen documents filled with sacred hieroglyphic symbols and dot-and-bar mathematical notations, blazing in front of the doorway of one

A portion of the Maya zodiac in the Paris Codex
showing a pair of birds flanking a long-nosed serpent.
(Bibliothèque nationale de France)

of their newly built churches. The destruction was practically total, save for a few fragments likely snatched by a bystander, perhaps for their souvenir value. Badly eroded, the Paris Codex is named for the city in whose library it resides. It was redis- covered in the 1850s, abandoned among a pile of soot-covered papers in a chimney corner of the Bibliothèque nationale de France. On what remains of pages 23 and 24 are pictures of a number of animals dangling below a continuous band, their jaws clamped around sun symbols. Judging from the style of the rest of the codex and from the content of an adjacent page, the band represents the body of the two-headed sky serpent (*caan*). The parade of animals continues across the lower band. All in all, thirteen animals make up a Maya zodiac.

The most easily identifiable creatures are a rattlesnake— you can clearly see the rattle; tortoise; scorpion; a pair of birds —one is probably a vulture; and a serpent. Less certain are what look like a frog, deer, human skull, and peccary. Are these celestial creatures devouring the sun, moon, and planets that enter their zodiacal domain? Below the animal images ap- pears a table of Maya day names and numbers, with twenty- eight days separating each date entry. Altogether, a twenty- eight-day period is registered thirteen times, totaling 364 days along each horizontal row. As in the Chinese zodiac, the num- ber twenty-eight implies a lunar connection. The ancient Maya seem to have been using this table to chart the course of the moon among the stars, dividing the lunar calendar into thir- teen months, each with twenty-eight days, so that a lunar year would consist of 364 days. But there is some evidence that the parade of Maya zodiacal animals is not laid out in a linear fashion, as we might anticipate. Instead the constellations seem to be arranged in alternating pairs so that, for example, when the first one is positioned just above the horizon on the east, the second lies just above the west, some 160 degrees away, as if in dialogue with one another across the landscape. Two other renditions of the Maya zodiac, each indicating the position of the planet Venus, show up on the friezes of ceremonial buildings in Yucatán.

The Maya zodiac also appears in an exquisite mural painting adorning the vault of a room in the Palace of Bonampak in Chiapas, Mexico. It depicts a scene of surrender in the aftermath of a battle, and the accession to office of the victorious ruler. Two of the constellations among four oval cartouches positioned above the celestial serpent skyband are recognizable from the Paris Codex: a tortoise and a pair of peccaries, shown in the act of copulating, flank a humanoid spear-wielding figure and another holding a ceramic vessel. Maya hieroglyphs representing the planet Venus adorn all four cartouches. In the gruesome scene below the zodiacal frieze, a group of cowering warriors, recently defeated in battle, plead for their lives to the exquisitely adorned king of Bonampak, who stands, bigger than life, hovering over those whom he has just subdued. One victim holds up his bloody hands, fingernails torn off in punishment, as the decapitated head of a victim rolls down the adjacent stairway beside him. Conquest in war and accession to power are well-connected themes in Maya mural painting and sculpture. Interestingly, dates on carved monuments at Bonampak refer to specific first and last appearances of Venus—the real "star wars"?

We have no idea how the Maya prophesied the events in the 1,200-year-old painting, and the descendants of the ancient Maya have little knowledge of how their ancestors used the codices. To illuminate what astrologers may have done in the past, then, we must rely on ethnographic studies of contemporary Maya practices. In one such documented case, diviner and client sat on opposite sides of a candlelit table adorned with pots of aromatic incense and rows of seeds and crystals. As in the Athenian agora, the client asked the questions. Is this a good marriage? Does this illness have an owner? (What is the cause of my illness?) The shaman drew items out of his divining bag, analogous to an astrologer consulting the codex, and, speaking to nature, replied, "I am now borrowing the breath of this day." Then he turned to each of the four directions: "I am borrowing the breath, the cold, the wind, the cloud, the mist at the rising sun (east), at the setting sun (west), at the four corners of the sky (south), and at the four corners of the earth (north)." Next he summoned the lightning in his blood so that it might speak truth to him. Today most divinations proceed by counting arrangements of crystals and seeds—removing and replacing the different piles, which represent various days and places where ritual offerings are to be carried out.

You may be disturbed—especially if you are scientifically educated—by these descriptions of how astrology worked for indigenous true believers. You may ask: How could believers trust their destinies to the divinations of someone so removed from the concerns of daily life? What if people in high places practiced the art today? Think of what could happen to the world! For most of us, astrology's tenets seem too illogical to trust, its actions too arbitrary and subjective.

Modern scientists have discredited astrology on the grounds that it has failed to synthesize its tenets into a comprehensive whole that can reduce and explain the movements of all the planets around the sun. Although we seem quick to praise the old Babylonians for their legacy of orderly notebooks filled with carefully acquired celestial observations, and the Maya astronomers for their precise, mathematically detailed codices, we accord little value to their cosmic ideas, or to the relationship of their beliefs about the sky to religious practice and daily life.

But as we wonder why these ancient peoples weren't more like us, we need to remind ourselves that, whether Greek, Chinese, or Maya, astrology's message was intended for people living in very different circumstances than ours. We mustn't think of astrological portents and omens only as fatalistic, unreliable predictions. Instead they served as prompts that pressed people to consider and talk about human affairs. The information sought by these ancient peoples is not emphasized in our scientific astronomy, but it felt essential to their lives, and they integrated this knowledge with deeply held beliefs concerning the relationship between nature and humanity. If we open our eyes wide enough, the zodiacs of the world can teach us much about ourselves.

4
Milky Way Sagas

The Maori of New Zealand tell a story about a river that circles both earth and sky, and a great warrior, Tama Rereti, who lived at the south end of Lake Taupo. In these early days there were no stars and it was so dark at night that people couldn't see their way around. Taniwha, a sharklike beast with huge spines along its back and large eyes that enabled it to see in the dark, would often attack and eat them. Fortunately it slept during the day in caves on the bottoms of lakes and deep rivers.

One day Tama Rereti awoke feeling hungry, so he decided to go fishing for his breakfast. He collected his fishing gear, placed it in his canoe, and pushed off into the lake. Taking advantage of the southerly breeze, he hoisted a sail and paddled to his favorite spot, about a half-hour boat ride from shore, where he caught three large fish. By the time he decided to head back to shore, however, the wind had picked up, so he resolved to wait out the wind by stretching out in the bottom of his canoe for a little nap. When he awoke, the young warrior found himself at the north end of the lake—and very hungry. He pulled his rig ashore, made a small campfire, and spit-roasted his succulent catch. When he glanced at his long shadow he realized it was getting close to sunset. There wasn't enough time before darkness to get back home on the opposite shore. Worse still, the time was fast approaching when Taniwha would surface and head out in search of a meal.

As he sat on a log by the shore and pondered his dilemma, Tama Rereti noticed the wet pebbles at the shoreline sparkling in the light of the setting sun. This gave him an idea. He loaded as many of them as his canoe could carry and pushed off into the lake, sailing to the place where the great Waikato River flows out of the lake and into the sky to create the rain. Following the swift current into heaven, Tama Rereti began to cast out the pebbles. The wake of the canoe became the Milky Way, and the pebbles became stars to light his way. Just as dawn broke and he tossed out the last of the pebbles, Tama Rereti spotted his village in the hills to the east, where the river descended. Meanwhile, Ranginui, Father of the Sky, was delighted with the beauty that Tama Rereti had created, and the aid he'd provided for people to find their way around at night. He asked the warrior if he would mind having his canoe placed among the stars as a tribute to his marvelous creation. With Tama Rereti's consent, Ranginui marked the decorated woodwork above its bow with stars in the head of Scorpius. Antares marks the crest of a wave where the bow meets the sea, with its elaborate carved post rising above the water. The stars in the tail of Scorpius are the stern. A bright star cloud in the Milky Way traces out the sail. The anchor rope follows a winding path of stars to Alpha and Beta Centauri ending in the Southern Cross, the anchor, which keeps Tama Rereti's canoe in place in the swift current of the southern Milky Way.

Where the Maori see a Milky Way made of glittering pebbles, the Hindus see a school of swimming dolphins, the

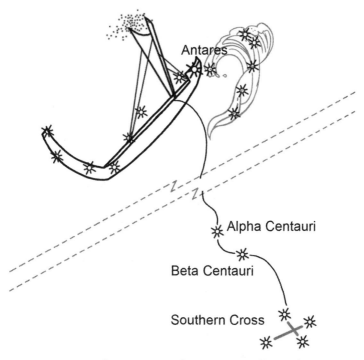

Antares

Alpha Centauri

Beta Centauri

Southern Cross

Tama Rereti's Canoe, an indigenous constellation from
New Zealand. (Drawing by Julia Meyerson)

Finns a flock of flying birds. Armenians imagine a thief who
stole a bale of hay, losing some of the dried grass during his
escape; for the Cherokee, the thief is a dog, spilling a bag of
cornmeal. Hungarians liken the Milky Way to horseshoe sparks
from cavalry hastening over the pavement to battle. Zulu people think of it as a cow's stomach, while the ancient Greeks saw
milk, sprinkled across the sky when the infant Heracles sucked
too hard on his mother's nipple. In modern times, we call the
Milky Way—that luminous band of faint light interspersed with
dark clouds that encircles the sky—our galactic home, one that

we share with two hundred billion resident suns and millions of solar systems spread over its hundred billion square light years of space. The Milky Way only looks like a roadway paved with stardust because we are wrapped in its wafer-thin, disk-like, spiral structure. The stars that populate the Galaxy are so far away that they blend together into a continuous background.

Our ancestors didn't always view the Milky Way Galaxy this way. Wise Aristotle thought it was swamp gas wafted up by the breeze into the sky. It wasn't until well after the invention of the telescope that observers resolved the diffuse light into stars and only in the 1920s did astronomers come to understand "our home" in the realm of the nebulae (to borrow the words of astronomer Edwin Hubble, discoverer of the expanding universe) to be one among billions of other spiral- and elliptical-shaped wheels that make up the universe at large.

Viewers in the Northern Hemisphere can most easily walk their eyes along the lighted pathway on late summer evenings, when its brightest portion crosses just south of overhead and aligns north-south. The Milky Way runs the length of the Northern Cross in Cygnus, passes southward through Aquila, then widens and brightens as you enter Sagittarius where, from our eccentric position, the sight line takes in a 25,000-light-year stretch toward the center of the lens-shaped stellar disk. The galactic bulge contains three-quarters of the Galaxy's mass, including the interstellar gas and dust out of which stars are born. If you could find a cosmic fan big enough to blow all the dust away, you'd be able to read a book by the light of the Milky Way.

The galactic center is a menacing way station along the bright path. It houses a supermassive black hole along with an abundance of exploding supernovae. Clearly, the volatile neighborhood that includes our home sweet home is not prime real estate for safely raising a family. Next the Milky Way crosses Scorpius and, if you can see far enough south from locations in the Northern Hemisphere, you'll be able to follow it through Centaurus and Crux, before it heads back north via star fields located between Canis Major and Minor, Gemini, and Orion. The Milky Way thins out and dims as it passes through Taurus and Auriga, where you're looking toward the galactic anti-center. Perseus and Cassiopeia complete the circuit.

Like the zodiac, the Milky Way is misaligned with the plane of the earth's rotation, so it also makes a kind of tumbling motion as it rises and sets. This motion is more extreme than the zodiac's because of its steep tilt (about 60 degrees compared to 23.5 degrees of the zodiac) in relation to the stars' motion over a twenty-four-hour period. Imagine lying flat on your back at dusk in early autumn with your head pointing south. Then you'll see the Milky Way ranging from northeast to southeast with a wide stretch of it passing overhead. By midnight it turns from east to west and at dawn from northwest to southeast. If you continue watching from night to night, you'll notice that the positions will shift; by early spring they reverse themselves. There are also times—for example, around midnight in early spring—when the Milky Way lies almost flat, passing all the way around the horizon. To appreciate the complex movement of the Milky Way, you really need to watch it year round. Viewing through the seasons, you could compare it to a bottom's-eye view of the rim of a coin spun around on a glass tabletop. No wonder the ancient Maya thought of it as the twisted umbilical cord connecting heaven and the underworld to the earth. Sadly the bright lights of cities and towns where most of us live deprive us of the view they beheld in dark Yucatecan skies.

Some contemporary Maya groups regard the Milky Way as a great celestial roadway. The Chorti Maya, for example, call it the Camino de Santiago, or the Way of Saint James—the network of paths in Spain that pilgrims take to reach the shrine of the apostle Saint James. They pay particular attention to its alignment in the sky relative to the position of the sun. Other Maya groups describe the Milky Way as a celestial river that carries the canoe of the paddler gods and First Father, our creator, into and out of the underworld, Xibalba. A design pattern in Maya art and sculpture that comes from the *Popol Vuh,* or Book of Counsel, features the Milky Way in its creation story. The zodiac is depicted as a two-headed serpent and the constellations positioned at one of the locations where it crosses the Milky Way are identified as especially important; that's where Orion, Taurus, and Gemini meet. The action begins with the lighting of the Three Stone Hearth, which, you'll recall, is the lower region of Orion including the Orion Nebula, by First Father, who was reborn out of the shell of a cosmic tortoise—the one sketched out by the Belt of Orion. He raises the great World Tree (another Milky Way designation), which first takes the shape of a crocodile. The Pleiades represent the handful of seeds that, once planted, will grow into the fertile World Tree, the portion of the Milky Way arising from the cosmic fire that passes prominently through the north-south overhead zone a few hours after the Pleiades have crossed the zenith. Each year the creation clock is rewound and the story retold.

Archaeological findings of ancient Maya writing and sculpture bolster the sky genesis story. Bone carvings from the ruins of Tikal depict a pair of gods, Jaguar Paddler and Sting Ray Paddler, as described in the *Popol Vuh.* They are shown guiding their canoe over the Milky Way en route to the place of creation, transporting their passenger, the sprouting Young

Maize God. As the Milky Way rotates from its erect north-south orientation in the guise of the World Tree, to its east-west position lower on the horizon, where it becomes the subterranean Cosmic Monster, or crocodile, the story of the world's creation plays out for viewers on earth.

In the high Andes, the Milky Way is all about the flow of water. In this precipitous environment, where elevations drop from 15,000 feet down to sea level over a mere hundred miles of terrain, it makes sense that people would be especially aware of the movement of this most precious liquid, which can disappear in an instant after a heavy rain. When will water come down from the heavens, which way will it go, and how can it be best used to nurture our crops?

In ancient times, the Inca say, the creator storm god Viracocha ("Sea Foam" in the still-spoken Quechua language) rose from Lake Titicaca (Bolivia). He crossed over the sky and entered the sea (off the coast of Ecuador). Contemporary residents of the village of Misminay, near the Inca capital of Cuzco (Peru), say that the Vilcanota River, a major carrier of water toward the sea, leads back up to the sky in the form of the Milky Way, or Mayu, at a point on the horizon where the river intersects it. The celestial river lines up in the same direction as the flow of the Vilcanota. As Spanish Jesuit missionary and writer Bernabé Cobo tells it:

> They say, in addition, that through the center of the sky there crosses a great river which they take to be that white band which we see from here below and call the Milky Way . . . Of this river, they believe that that it takes up the water which flows beyond the earth.

In contemporary Misminay, the Milky Way is the most important visual aid for orienting celestial with terrestrial space. On
early evenings during the dry season, when the Milky Way
first becomes visible in Misminay, it stretches from the northeast to the southwest. During the rainy season, the path runs
southeast to northwest, and aligns with the Vilcanota River at
either end, sending the heavenly water back to earth.

Residents connect time to space by marking the four seasonal dates when certain *chaska,* or bright stars, located near
the Milky Way, appear and disappear. Pilgrimages following the
Milky Way along the rainy season axis have been traced all the
way back to precontact times, when Inca priests walked from
Cuzco to the wellspring of the Vilcanota along that path. They
walk the terrestrial Milky Way to make offerings to the gods.
Every June more than fifty thousand people, including some
tourists, still participate in the Q'olloy Rit'i pilgrimage to the
21,000-foot-high Ausangate mountain in the Cordillera Vilcanota range, where melting glaciers still serve as a major con-
tributor to the delicate Andean watershed.

Descending eastward from the Andean highlands into
the rainforested Amazon, the theme of sacred water persists.
Members of the Barasana tribe, who hunt, fish, and gather in
the northwest Amazon basin, tell outsiders that they live at the
center of the world. They make a good case for it: since they
reside on the equator, they see the stars move along vertical
daily paths on either side of the east-west line and observe the
sun passing straight overhead on the first days of spring and
fall. They call the stars the Universe People (*umuari masa*). According to Barasana beliefs, they were created by the primal
sun as his children, and when they died they were revived and
granted immortality, the next step in the creative process, in a
space-time world beneath the earth that is the opposite of our

mortal world. When we bask in daylight, their world is dark; our rivers flow east-west, theirs west-east. Their behavior unites men with women, earth with water, and so on in a series of complementary dualities.

The most important Barasana Universe People live along the Milky Way, or Star Path (*nyokoa ma*), which they divide into two segments, somewhat like their Andean neighbors to the west. There's the New Path, which extends southeast to northwest, and the Old Path, which aligns northeast to southwest. They list their Milky Way inhabitants, ten to each segment, the way we line up our parade of zodiacal animals. Most important are the lead constellations: the Pleiades heads the New Path; Scorpius, the Old Path.

The Pleiades, or Star Thing, is the Woman Shaman. Her stars are composed of the wooden strips she used to set fire to the forest, clearing it to create the first fields. When Star Thing appears in November, she heralds the end of the rains and the beginning of forest clearing by men. When the New Path residents of the Milky Way light up the night sky, they cue the Barasana on subsistence duties they need to tend to and the edible treats that will ultimately reward their helpful behavior. Among the good things ushered in by New Path constellations are the Fish Smoking Rack (the Hyades), the Adze (Orion's Belt and Sword), Jacunda Fish (the region around Rigel), and Crayfish (Leo).

But beware the illumination that descends during Old Path's time of the year, which brings with it warnings about dangers that can lurk in the dead-dry season: the Scorpion (our Scorpius, Lupus, and Libra), Poisonous Snake (Corona Australis), Vulture (Aquila), and the Corpse Bundle (Delphinus) of a Star Woman stung to death by a swarm of bees. They say she fell to earth as a shooting star, then came back to life, married a mortal, and returned to the sky—only to be bitten and killed a second time by a star snake.

At dusk in mid-November, Barasana men, women, and children sit in family groups outside their cluster of huts and gaze at the Milky Way. The skywatchers focus their attention on the Star Paths at opposite ends of the horizon: Star Thing rising in the east in front of them and Caterpillar Jaguar setting over their shoulders. They pay attention to the vertical ascent of the eastern stars because their act of rising lifts away the rains. They see the Milky Way as the Milk River, a continuation in the sky of the Amazon River here on earth. Where the sky meets the earth on the east side there is a great waterfall that sends precious liquid to the world below and an underworld river that brings it back again. They see it return in the water that flows down the sides of mountains on the opposite side. Water circulates in a closed loop through the entire cosmos.

Not content merely to watch passively, the Barasana dance to the Milky Way in order to urge the stars along their courses. They form two lines of ten and dance around a central area they call the center of the world. One moves left to right while the other circulates right to left, replicating the motion of the stars. They continue all night long, or at least until the stars begin to fade from view in the morning twilight.

The natural world of the Barasana is teeming with life, both terrestrial and cosmic, in a way that few outsiders can fathom. The biodiversity in the tropics is extraordinary: 90 percent of all living species thrive in rainforests where the Barasana make their home. And Barasana belief systems, based in part on their skywatching practices, support the idea that human action helps the world go round.

Yet in many ways, these belief systems are completely understandable. For example, the concept that the sky over the place where we live is a solid dome, inhabited by other people who may be within reach, isn't so different from the Western logic that allows for the possibility of extraterrestrial life in astronomers' "vast sea of habitable worlds." "I will make a road for the people to travel along when this thing happens," declares the Ojibwa creator god Manabozho to his people. The "thing" is death, the mystery that animates all humankind. Why does it happen? What happens next? Where do we go? How do we get there? According to Ojibwa legend, Manabozho's brother was the first being to die, drowned by the Spirits of Water, and that's how death was brought into the world.

Manabozho laid out his answers about death to the frightened listeners:

> This is what the people will do when this thing happens to them. [Next] he went toward the sunset. As he went along, he made four signs of places. He put four [spirit powers] along the way . . . Otter on the right side, owl on the left, hills (snakes) on both sides, a river with a snake/log . . . Then the road forks into a short path, which is bad and forever, and the one that continues on behind the sky, beyond the sunset.

Once you close your eyes for the final time, your shadow will leave your body and take this road. You'll know you're going the right way because you'll meet signs. There will be a dark windblown tunnel, then an old woman, Our Grandmother, who will direct you to four old men, "Our Grandfathers." They will tell you where to cross the half-red, half-blue river. You'll find a huge log near the shore that you can use for a bridge. On the other side you'll be given further instructions on how to climb up to the Tchipai Meskenau, the Path of Souls of the Milky Way.

The Cherokee say the entryway to the Path of Souls is guarded by two dog stars, Sirius and Antares, where the pathway is accessed on opposite sides of the horizon; that's where the Milky Way meets our world. But you'd better be sure you've brought along enough food to offer them, or they won't allow you to pass. (Like the Lakota Hand constellation story, this is a lesson to remind you that you must always remember to make your just offerings to the gods when you are alive, or things might not go so well once you're dead.) Most important of all, you will need to time your trip carefully so you won't miss the doorway that gives you access to the Milky Way. If you jump too early or too late, you'll fall into the water and land in the Beneath World. The window of opportunity is available only for a few minutes each night, when the Milky Way lies low in the sky, nearly parallel to the horizon.

If you look closely at African art, particularly carved sculptures of ancestral figures, you'll notice decorative lines that run from the nape of the neck down to the navel on one side and down the backbone on the other side. You can see the same pattern in the line of scarification along the backbones of members initiated into healing societies. It also appears vertically down the belly during pregnancy. They call it the *mulalambo,* the line that divides the universe in two, and creates the harmonious balance of left and right.

A native Tabwa ethnographer tells us that his people, who live in eastern Congo along the grassy shores of Lake Tanganyika, think of the sky as a magnificent work of architecture, a solid vault supported at the ends of the earth by copper columns, the same way the roof of their round house is held in place. The mulalambo that divides the sky is the Milky Way, an extension of the symmetry that also partitions their great lake. It splits into two zones where the winds meet. They say that different kinds of fish are caught on either side. Those along the terrestrial mulalambo are bigger, fiercer, more aggressive, because there's a zone of conflict along the line of symmetry. The Milky Way is a pathway wandering across the center of the sky that completes the circle of the universe. It is also part of the Pathway of God followed over both land and sky by the hero Kyomba, who first led his people to Lake Tanganyika's shores. With the advent of technology, the cosmic mulalambo has also been likened to a timepiece. Tabwa hunters call the Milky Way the bow of the night because of the way it curves over the sky from horizon to horizon. They say it is "our clock," and they tell the time of night by the degree to which its "hands" have turned, the way the hands move on the wristwatches they wear.

Members of the ancient Tang dynasty thought of the Milky Way as a waterway that complemented the rivers of their kingdom, the Rivers Ho (the Yellow River) and Han. You could best access it between the retreat of the summer monsoons and the advent of the dusty winter months; that is, during early autumn, when the shining sky waters are unruffled by windy waves.

A Tang poem tells the story of the annual arrival in late summer of an unoccupied raft on the shore near the home of a man who lived by the River Ho. It would stay for just two days, then sail away. The man was often tempted to hop on the craft to see where it might take him, but year after year he demurred. One day, "possessed by a strange ambition," he boarded the craft, taking along an ample supply of provisions, and took off with it. At the edge of the horizon he floated up to the Sky Ho and journeyed for ten days among the sun, moon, and planets. Gradually they disappeared behind him, and he could no longer tell whether it was day or night. Finally, even the starry background faded from view. On he floated for another ten days until he reached the shore of what looked like a civilized place. He could see houses inside city walls, fortifications, and many large impressive buildings—a huge city in the sky. Eagerly stepping off the raft, he encountered a man who had stopped by the river to water his cow. The man seemed startled and asked him, "What are you doing here? From which place did you come?" Our man told him that he was curious and had wanted to make the trip over the Milky Way to see what other civilizations might exist. Then he asked, "What place might this be?" The reply was immediate and curt: "Go back where you came from. There is nothing here for you."

In the end the cosmic voyager never went ashore. Instead the raft returned him home safely. Today they still tell the story here on earth that there once was a "Stranger Star" who trespassed on the Milky Way constellation known as "Ox Hauler." (In ancient Chinese astronomy, the word *trespass* conveys planetary invasions of constellations—unknown cosmic territories where people like us are not welcome.)

With its opposing pathways that travel back and forth along the horizon nightly and seasonally, even dropping down flat to the level we inhabit, the Milky Way seems so accessible. Appearing as a celestial continuation of terrestrial rivers or roadways, it has long beckoned believers to enter the route to heaven—a call that has been heard, and answered, by countless peoples since human cultures have emerged. Humans will undoubtedly keep responding to this great display of our enormous home galaxy. Like the ambitious Chinese voyager's search for supernal bliss, we will continue to venture into the cold, vast darkness of space in search of answers to our own age-old questions.

5

Dark Cloud Constellations of the Milky Way

The leader of the Australian Wotjobaluk was growing tired of the Tohingal, a giant emu that terrorized villages and ate people. He decided to team up with the powerful Brothers Bram, a pair of white cockatoos, to kill the beast. The three crept up on the emu's nest and tried to surprise Tohingal. The big bird awoke and charged, splaying out his huge feet to crush the attackers, but the brothers quickly launched a one-two blow, delivering a spear first to the bird's neck and then another to its rump. Fatally wounded, Tohingal escaped, losing blood as it staggered onto the northern plains. They say its stream of blood became the Wimmera River. If you live in the Southern Hemisphere, you can see the defeated Tohingal, crumpled in a heap, in the dark gaps of an otherwise bright Milky Way.

The Alpha star in the Southern Cross shows the spear point destined for the emu's neck, while Beta is the one headed for the giant bird's backside. The top star in the Cross is an opossum that was chased up a tree by the emu during the fracas, where it remains, ever the nocturnal dweller. And the Brothers Bram? They are the bright stars Alpha and Beta Centauri. Tohingal's conquerors separated the feathers and split each one down the middle, carefully placing them in two separate piles. One became the males, the other the females, of the emus descended from the giant creature. Look closely and you'll see the sharp dividing line down every emu feather, a reminder of their origin.

The Kamilaroi and Euahlayi tribes of New South Wales don't know Tohingal. Instead these tribes remember a blind man who lived with his wife in a camp in the bush. Every day she needed to go out and hunt for emu eggs. No matter how many she brought back, the demanding husband would always complain that they were too small. One day, she spotted some unusually large emu tracks and followed them to a nest filled with gigantic eggs, tended by a huge male. When she tried throwing stones at the emu to get him off the nest, the emu charged her and kicked her to death with his muscular three-toed feet.

When his wife didn't return, the hungry blind man became concerned. He groped around and managed to grasp a bush laden with ripe berries. When he ate the berries, his sight miraculously returned. Picking up a sheaf of spears, he set off to find his missing mate. When he came upon the giant emu standing over the corpse of his trampled wife, he immediately attacked it with one of his spears and sent it into the sky, where everyone can witness the giant bird's menacing shadowy figure silhouetted against the milky light.

Aboriginal Australians give the name "Dreamtime" to the era when ancestral figures created our present world. They call their tales "dreaming stories" and believe that these stories give meaning to special places and creatures that are still with us.

The "Emu in the Sky," a constellation from Aboriginal Australia,
is comprised of shapes formed by the darker areas of the
southern Milky Way. (Ray Norris and Barnaby Norris)

Dreamtime stories like the one of the emu in the sky are among the most cherished. The Australian celestial emu stretches over a quarter of the southern sky, from the head of the beast, in a dark spot called the Coalsack, through the Southern Cross, then across the dense dark region of the Milky Way to Scorpius, which outlines the body, and Sagittarius, where a bright star cloud represents the eggs it sits on.

The seasonal movement of the emu in the sky animates the Dreamtime myth. Being close to the south celestial pole, the celestial emu's head is visible from Australia every night, but the entirety of the giant bird doesn't appear until April and May. That's when, people say, it runs across the night sky, in concert with the mating season among emus, when the females frantically chase the males. The emu's posture also signals that the tasty eggs will soon become available. In July the emu's legs disappear below the horizon and the cosmic bird is transformed into a male that sits on the nest, brooding. Significantly, the bright star clumps just north of what astronomers call the Trifid Nebula in Sagittarius look like eggs, and September is the last time the eggs are available to gatherers. At this point, the head and neck of the giant emu in the sky join the feet "underground," leaving only the body of the emu on the horizon, now transformed into an egg. The Kamilaroi and Euahlayi schedule their male initiation ceremony just after the eggs hatch. As the male hatches the emu eggs into chicks, so the elders welcome the male initiates into the adult world.

The Milky Way is more than a band of bright streaks of diffuse starlight. Woven among the carpets of stars are dark lanes and blotches once thought by astronomers to be empty space between the stars. It wasn't until the beginning of the twentieth century that they realized Aristotle was correct: nature really *does* abhor a vacuum. Handfuls of faint stars visible in these darkened pockets looked excessively red, as if they were hiding behind or embedded in a medium that altered their color (much as the way the sun looks redder when it rises or sets because of scattering by tiny particles in the atmosphere). Further studies of starlight passing through the Milky Way showed an abundance of hydrogen and simple molecular gases—the stuff that gives birth to the stars—in the interstellar matter within its spiral arms.

If you follow the glowing road in the sky beginning at the Milky Way's busy intersection with the zodiac in the constellation of Taurus, and pass through Perseus and Cassiopeia, you'll notice a few dark patches in the luminous background; as your eye journeys southward, the dark clouds become more conspicuous, their edges more well defined. They have even been given names, like the "Great Rift," also called the "Dark River," and the "Northern Coalsack," a dusty lane that stretches from Cygnus to Aquila and on to Sagittarius. The Great Rift contains more than a billion earth masses' worth of interstellar gas and dust.

Past Scorpius, well out of range of most observers in the Northern Hemisphere, lies the most prominent of the dark clouds in the Milky Way, the Coalsack. About forty times the size of the full moon, it covers the constellation of the Southern Cross and a portion of Centaurus. Westerners probably wouldn't call them constellations, but the Great Rift and the Coalsack were among a host of dark images along the Milky Way that were given diverse names and meanings by storytelling inhabitants of the Southern Hemisphere, especially Australian and South American skywatchers.

The sixteenth-century historian Felipe Guamán Poma de Ayala, who was half-Inca, half-Spanish, wrote this about the Coalsack and its vicinity:

> They fancied they saw the figure of an ewe with the body complete suckling a lamb in some dark patches spread over what the astrologers call the milky way. They tried to point it out to me saying: "Don't you see the head of the ewe? There is the lamb's head sucking; there are their bodies and their legs." But I could see nothing but the spots, which must have been for want of imagination on my part.

The ewe is actually a llama, a camel without humps. Indispensable in the Andean highlands, they are highly socialized, intelligent animals who live in herds, usually tended by native women. Llamas are capable of carrying up to one-third of their body weight and the wool they produce is very soft, woven to make heavy garments and *quipu*, knotted string devices employed by the ancient Inca for recordkeeping.

Yacana, the celestial llama, is the animator, the maker of llamas on earth: "We native people can see it standing out as a black spot," a native told another chronicler. "The Yacana moves inside the Milky Way. It's big, really big. It becomes blacker as it approaches through the sky, with two eyes and a very large neck . . . [It] has a calf. It looks just as if the calf were suckling." Look closely and you'll notice that the baby llama (*uñallama-cha*) is still attached to its mother by an umbilical cord. The mother's body is formed by the Coalsack, near the Southern Cross. Her glowing eyes are the bright stars Alpha and Beta Centauri.

The movement of the llama in the sky is intimately tied

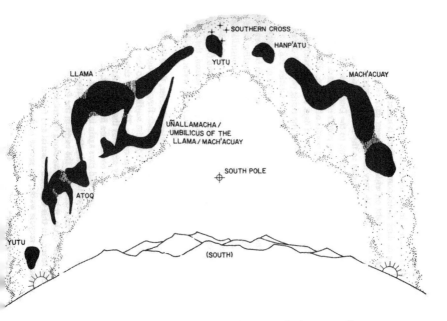

The Andean llama dark cloud constellation with the constellation
of the predatory Fox (*atoq*) nearby. (From *At the Crossroads of
the Earth and the Sky: An Andean Cosmology,* by Gary Urton,
copyright © 1981. Courtesy of the University of Texas Press.)

to the agricultural calendar in highland Peru. Each month of
the year, the Inca sacrificed one hundred llamas of different
colors (white, brown, and multicolored) from among the great
numbers of llamas brought to the Inca capital of Cuzco for this
ritual. One chronicler tells us that in September (the month
of planting), large numbers of animals due to be sacrificed in
the next year were brought in from the fields, where they had
grazed since the previous year's harvest. Another schedule of
llama sacrifices was timed to coincide both with the celestial
llama's disappearance and reappearance, as well as with the
dates when it stood at its highest and lowest sky positions at

twilight. Viewed from Cuzco, the rising and setting points of the llama's eyes are neatly framed by the boundary lines of the southernmost quarter of the city.

To better understand what the moving image of the celestial llama meant—and still means—to the Andean people, let's think for a moment about how the movement of things in the sky satisfies *our* needs. Our seasonal calendar consists of 365.24219 days, determined from careful observations that measure how many earth rotation periods of twenty-three hours, fifty-six minutes, and four seconds (the day) fit into the time it takes the earth to revolve around the sun (the year), which we experience as the time it takes the sun to pass all the way around the zodiac. We can measure it to an accuracy of better than a second per year. Occasionally we insert a leap second to make up for the earth's spin slowing down. Living in a fast-moving techno-world we have developed a penchant for precision! But our needs have changed through history. Living in the low-tech world five hundred years ago, the ancestors of the modern West were content with clocks that had only an hour hand. In the days before the Roman Empire, time was kept on sundials that meted out unequal hours, and it wasn't necessary to keep track of the entire 365-day seasonal cycle. Farmers needed to know only the months that mattered, that is, the ones that spanned planting through the end of the harvest —a period they tallied by full moons. Later pre-Roman ancestors reckoned a 305-day (ten-month) year, which they linked to the gestation period of cattle. This made sense because they lived by the life cycle of their most important domesticated animal: the one whose flesh nurtured them, whose backs bore their burden, whose hair provided thread for their weaving, whose hides warmed their bodies, and whose bones offered material for tool making.

Similarly, the year of the Inca lasted 328 days, the interval between conception and birth in the life cycle of the llama. Thanks to a bio-astronomical coincidence, they measured it by subtracting the time between the last disappearance and first reappearance of the Pleiades, thirty-seven days, from the full seasonal year of 365 days.

The llama sky myth also tells us about Andean irrigation practices. According to the Huarochirí Manuscript, a sixteenth-century document from highland Peru annotated by the Spanish cleric Francisco de Ávila:

> The *Yacana* [the celestial llama] ... is like the shadow of the llama. They say that this *Yacana* comes down to earth at midnight, when nobody is aware of it, and drinks all the water from the sea. They say that if she did not drink this water the entire world would be drowned.

There follows an omen that tells what happens when the llama drinks:

> They say if a man was in luck and fortunate, the *Yacana* would fall right on top of him while it drank water from some spring.
> As its woolly bulk pressed down upon him, someone else would pluck out some of its wool.
> That apparition would occur at night.
> In the morning, at daybreak, the man would look at the wool he'd plucked out. Examining it he'd see the wool to be blue, white, black, and brown, of every hue, thickly matted together.
> If he had no llamas, he'd worship at the place

where he had seen the apparition and plucked the wool, and trade for some llamas right away. After worshiping he'd trade for a female and a male llama.

Just from the two he'd bought, two or three thousand llamas would soon come.

In old times the *Yacana* revealed itself this way to a whole lot of people all over this province.

Andean farmers still pay their debt to the animator llama. Llamas of the different colors mentioned in the story are sacrificed at specific times: brown and brownish-red at the beginning of planting, black ones tied to posts and starved to induce rain and crop growth in midseason, and multicolored ones at harvest time.

So exactly when does the llama drink the water? Her eyes disappear after sunset in mid-October as her head plunges into the horizon to take a drink from the swollen rivers that threaten to drown the world. She returns, rising in the east a little over a month later, bringing with her a signal that she must be shorn of her wool. This is a dangerous time when Andean pastoralists need to pay close attention to newborn calves. They must be aware of predators, especially the fox. If you look into the sky, near the suckling calf, you can see the crafty fox—that is, the dark cloud constellation Atoq—following closely behind the mother, at a right angle to Scorpius's tail pointing toward Sagittarius. Mother Yacana seems to be using her powerful back legs to trample the threatening Atoq.

Other dark cloud animals seem to float along the Andean sky river. They include a pair of tinamou, or partridge (*yutu*), flanking either side of the llama-fox drama, a toad (*hanp'atu*), and a giant anaconda (*mach'acuay*) to the west. Their life cycles on earth are also clocked in the sky. For example, the celestial toad rises in the east in the morning, just after terrestrial toads end their period of hibernation and begin croaking—the louder the better to secure a mate. The celestial snake, too, is made up of a long dark cloud that streaks from the Southern Cross to the west side of Canis Major. It rises headfirst, the way snakes reenter the world in the warm rainy season when they first come out of the ground, and returns to the earth at the beginning of the cold dry season, when the terrestrial snakes also stay underground. The bio- and astro-cycles of animals above and below have always mimicked one another.

Most effective stories are adaptable, so that a skilled teller can use "mythic substitution"—in other words, he or she can alter its characters and scenarios to appeal to various audiences. The highland Andean myth of the dark cloud llama and her baby being stalked by a fox is a good example. When the story traveled eastward into the tropical rainforest of the Amazon basin, the carnivore predator fox became a jaguar who chases the herbivore prey, a tapir, down the Milky Way. In the foothills between the two zones, the tapir is replaced by a deer. Farther south, in the Gran Chaco grasslands of southern Chile and Argentina, the constellation pair becomes a dog and a rhea, or South American ostrich; only this time, because of the ostrich's long neck (much like a llama's), pursuer and pursued switch positions in the sky.

Not surprisingly, all stories that involve dark cloud constellations emanate from Southern Hemisphere cultures. The Milky Way over the Amazon is envisioned as a battlefield between two indigenous animals. You wouldn't think a long-nosed, bushy-tailed giant anteater would have much of a chance coping with a fierce jaguar on the prowl, but modern documentarians have reported that in those rare instances when they have witnessed an encounter between the two in the South American rainforest, it often ends in a draw. Extremely sharp claws, normally used to dig insects out of the ground, help in the anteater's defense.

Such confrontations are quite familiar to the Shipibo, who live along the Ucayali River in the western Amazon basin, where the anteater, like the coyote among the Navajo, plays the role of a trickster who tries to foil the jaguar. The Shipibo tell the story of the long-nosed one issuing a challenge to the jaguar to compete in an underwater breath-holding contest. The jaguar accepts, so both remove their pelts, lay them on the shore of the river, and dive into the water. Suddenly the crafty anteater jumps out of the water and steals the jaguar's pelt, leaving the victim of his deceit with his own discarded coat. They say that ever since, the two creatures have remained trapped in each other's skins.

Look up at the dark spots in the Milky Way, toward the South Pole, and you can see the two animals tussling. The anteater's body is made up of the dark Coalsack, while the jaguar is the bright patch adjacent to it on the north. As they rotate across the sky locked in battle, the anteater gains the advantage by appearing on top shortly after sunset. Later in the night, however, the positions become reversed, with the jaguar gaining ascendancy as dawn arrives. Next night the fight resumes. The stark meeting between light and dark areas of the Milky Way symbolizes transformation and the identity crisis that can follow, a dilemma we all face sometime in our lives.

Like the dark-cloud Australian emu and Andean llama, the shadowy constellations observed by the Desana of Colombia also mark critical times in the life cycles of animals that matter to them. For example, they make the Great Rift out to be canoes filled with caterpillars that navigate across the night

sky, descending to earth on the east horizon—where the winds
pick them up and carry them across the land on long threads
of silvery saliva. The Desana caterpillar shamans become very
stressed when these creatures emerge from the sky. Their
thoughts and feelings turn to whether their appearance por-
tends danger. As professionals they need to make the proper
incantation to mark the swarm's appearance. Speaking to an
anthropologist, one of them anxiously reacted to the coming
of the caterpillars as he attempted to assess the potential evils
that might endanger those attending the ceremony and pre-
scribe the appropriate narcotic drug to administer in order to
ease tensions over these potential threats:

> Well, there it is, isn't it? There, at the river mouth,
> yes. There are the heads of the *ii* caterpillars. Yes *ii*
> caterpillars with red heads, yes. When these heads
> appear we say they are coming in their canoes, and
> when the *ii* heads come, the winds come with them.
> The canoes come with a rushing sound, we say. The
> wind is like a torrent. Then those here take *vihó*
> snuff; they snuff it and absorb it, yes. And once
> more they take snuff. They mix it with the starch
> of *carayurú* . . . In our visions we then see a crowd
> of people who try to kill us. That's the way it is,
> elder brother. That's it. They say they have red heads.
> Then there are some with spiny heads, too. Those
> with spiny heads are evil. Then the *vihó* snuffers
> gather there. This is why it thunders. That is what
> they say; isn't it? But they come later, the spiny
> heads; the red heads come first. These *ii* heads are
> dangerous in their time; but not now. Now they do
> not make any noise. [But] then the windy currents

come, the wind of these canoes; don't they? This is why we say they are coming with the wind, and then there is danger . . . It is such that it produces nightmares in us . . . When they behave like this, in a crowd, they throw us into confusion; yes, elder brother. It makes us nauseated, the women and the canoes of the *ii* caterpillars. This is why we become nauseated. This is what they say when talking. That's all.

6

Polar Constellations

Walking on thin ice poses a hazard for the Arctic seal hunter, especially if that hunter is overweight. Once there was a huge man, named Sikuliaqsiujuittuq, who avoided the risk by stealing from successful hunters. He stole so often that he acquired a reputation. He learned to tell how successful the other hunters were by checking out their wrists when they returned from the hunt. If the wrists were dirty, he knew they likely had not been immersed in water, so these men hadn't caught anything. One day, when the sea ice had reached the same thickness as the land-fast ice, the other hunters persuaded the huge man to go out for the night and camp with them. "We usually sleep with our hands tied behind our back," they advised him. Since it was his first hunt, he didn't question the odd advice.

In the middle of the night the hunters tried to stab him, but Sikuliaqsiujuittuq was very strong and he managed to break the bonds. During the ensuing struggle, however, the big man's wound proved fatal. He ascended to the sky, where he became what astronomers call Procyon in Canis Minor. You can see the star's bloody reddish color when it lies low on the horizon, following the hunters in Orion's Belt. Later, two of the murderous hunters showed up at the huge man's dwelling, aiming to kill his wife and their two small children before she could seek revenge. But as they drew their knives, the woman kicked one of them to death, then grabbed the other by the throat and choked the life out of him. So ends the gripping tale of "The One Who Goes on Newly Formed Ice." Listeners will probably never forget the lesson, so important in hunter-gatherer societies, that it's important to pull your own load when it comes to food procurement.

Another star-based tale, a navigation riddle on the Arctic ice, takes place in midwinter. Two men pursued seals too far from shore and found themselves marooned on sea ice. One of them said, "I will follow the star Singuuriq" (Sirius). That will surely get me back to thicker, safer ice." Replied his companion, "I think it will be safer to follow Kingulliq (Vega), so that's what I will do." Which of the two made it home? Knowledge of the Northern Hemisphere sky gives the answer. In midwinter, Sirius appears above the southern horizon when Vega is visible low in the north. The hunter who headed south toward Sirius, where warmer water is located, was doomed by the more treacherous environment, while the one who followed Vega was destined for a more stable frozen surface. He survived. Navigation is tricky at the top of the world.

It's nighttime and you're standing at the North Pole (latitude 90 degrees when measured north of the equator). The imaginary axis of rotation of the earth runs up your spine and out the top of your head. It points to Polaris, the North Star, which marks the north celestial pole, the extension of the earth's geographic pole onto the sky, 90 degrees above the horizon. As the world turns and night progresses, the stars appear to move in the direction of opposite rotation along paths parallel to the

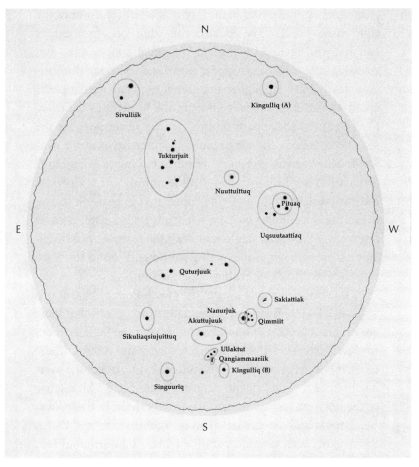

An Inuit star map of the north polar region.
(With permission of the Royal Ontario Museum © ROM)

horizon, each keeping a constant altitude above the horizon. If
you persist long enough, you'll see each star complete a portion of its twenty-four-hour circular orbit pivoted about the
North Star, with those stars farther from the celestial pivot

point circling in a wider path. Daylight won't return until the sun crosses the celestial equator, the extension of the earth's equator, which lies at the horizon. Eventually, around the first day of spring, you'll see it appear above and roll along the horizon, slowly climbing higher and higher each day, until it peaks about a quarter of the way up from the horizon to the overhead point. That happens on the first day of summer. As the season progresses, the sun gradually corkscrews its way downward, disappearing just after the start of autumn. Six months of daytime, six months of nighttime, minus a few weeks of morning and evening twilight sandwiched in between. If you live at the North Pole you experience one long day followed by one long night over a single year, with the same set of constellations, perpetually visible, going round and round the fixed celestial pole, 90 degrees above the horizon.

Of course, nobody lives at the North Pole; at least at this writing, it is embedded in a perpetually frozen ocean. But there are itinerant groups of hunters who inhabit the high latitudes scattered across northern Siberia, Canada, and Alaska, where the pivot of stellar movement is shifted slightly out of the overhead position. Consequently, star trails are tilted a bit relative to the horizon, so that some stars do set briefly. In these locations, night and day are less than six months long and there's a period between when, depending on how far north they live, people experience day and night sharing a lopsided twenty-four-hour schedule. For example, imagine you travel south from where you stood at the geographic pole. Walk over one degree of the earth's surface, approximately seventy miles, and Polaris will have moved one degree from the overhead point. Its altitude above the horizon would be 89 degrees, and you would be located at 89 degrees north latitude. Your *latitude* is the same as the *altitude* of the Pole Star (minus small corrections because Polaris isn't located precisely at the pole). Early navigators, like Columbus, put this principle to good use in finding their way across vast oceans, sailing along a chosen parallel of latitude by keeping Polaris at a constant altitude.

Given the tales told so far in this book, you might guess that anyone living under polar skies would dream up constellations and narratives about how to subsist in such a harsh climate: when and where to hunt before the long looming darkness, how to travel safely over the ice, how to go about acquiring sources of light and heat to endure the cold winter nights, and when to anticipate the nurturing sun's return after its lengthy absence. Among the handful of Arctic inhabitants we've asked, this is exactly what they've done.

The Inuit—which in their tongue simply means "people" —are a group of some 150,000 people, related by language, who live in the arctic regions of Canada, Alaska, Greenland, Russia, and Denmark. In Canada the Inuit occupy their own self-governing territory. They call it *Nunavut,* meaning "our land," and it comprises parts of northern Quebec, Labrador, and the Northwest Territories, centered on the capital of Igloolik at latitude 70 degrees north. There the sun is gone for two months, from late November to late January. From late May until late July, the sun shines continuously.

As we have seen, the Inuit sky is animated with tales of the hunt and the hazards of traveling over vast ice fields to secure the catch. Also, there's a bear, a caribou, and a seal they eat; foxes and wolves that threaten to snatch their prey; an oil lamp; a kayak stand; and, most important of all, a late winter harbinger of light. The Inuit don't have much use for Polaris. They call it Nuuttuittuq, or "never moves," and it's too high to navigate by. But they pay attention to the nearby constellation

Tukturjuit (Big Dipper), which reveals two images of caribou: one with the bowl and handle representing the body, and another where all of the stars represent individual caribou in a herd. A wolf, Sivulliik (the Boötes constellation), chases them around the pole.

"My father taught me about how to tell time and direction by following the big caribou through the night," related one unnamed elder. "When the caribou stands on its hind legs and its head starts to get higher . . . midnight is approaching." If you're out on the sea ice and the shore of Baffin Island is obscured, you should "raise your left hand and when your fingers and thumb match the stars with the thumb covering Phecda and the forefinger Megrez, the two stars opposite the pointer stars, and the remaining fingers the stars in the handle, your arm will point toward the mainland." Cassiopeia or Pituaq, located opposite the pole from the caribou, is variously represented as a seal, a sealskin, a container holding seal oil, or a seal oil lamp or lamp stand. Its three brightest stars map the three stones placed upright on the floor of a dwelling on which the lamp is placed. They say that all the lamps are sure to be lit when Cassiopeia stands level with the caribou.

A scene depicting the drama of the hunt looms high in the south during late evenings in the early months of the year. Nanurjuk, the polar bear (Aldebaran in Taurus), appears in the northeast at sunset. It circles all the way around the pole, disappearing in the northwest at dawn. Its cubs (the Hyades) stay close by, as a group of brothers and their dogs (the Pleiades) attempt to head it off. The hunters, or runners, Ullaktuit, are the belt stars of Orion that chase after Nanurjuk. "I have dropped my mitt!" yells one of the young hunters. "There's a full moon, no need to fear anything," responds an elder brother. "Go back and get it." As he turns, his brothers suddenly go up to the skies.

You can still see the brother who turned around and ended up stuck back on earth. He's the star Western stargazers call Rigel.

Of all the Inuit constellations, there's one you never could have predicted. Quturjuuk, the collarbone, extends in a line from Pollux and Castor in Gemini, to Capella and Beta in the constellation of Auriga. You can feel your collarbone, actually a pair of clavicles, by running your index and third finger from shoulder to shoulder; there's an indentation in it just below your chin, where the two attach to the sternum. It curves just like the line traced among those stars. Why a collarbone constellation? Anyone familiar with the delicate art of flensing, using a sharp knife to remove the fatty outer layer of skin from flesh, would be aware of how to navigate spaces between the skin and cartilage of a seal. Maybe this constellation was dreamed up by an imaginative Inuit butcher who became curious about similarities between the form of a seal's body and his own.

The sighting of Aagjuuk, the "stars seen at dawn," made up of the stars Altair and Tarazed at the top of our Aquila (the eagle), was the most important event in the Inuit annual dark-light cycle. These stars are the so-called bearers of light because they make their heliacal rise just before the dawn of the new year, a time when bearded seals migrate from the sea to the icy shores where they can be hunted more easily. The appearance of Aagjuuk also signals the start of the biggest celebration of the year. During the extended twilight, people eat and drink to excess; they masquerade and change partners—in other words, they become someone other than themselves. A disapproving eighteenth-century traveler who witnessed an extended Aagjuuk celebration wrote in his diary:

> They assemble together all over the country in large parties, and treat one another with the very best

they have. When they have eaten so much that they
are ready to burst, they rise up to play and to dance.
[A performer] expresses his joy at the return of the
sun in the hemisphere . . .
The welcome sun returns again,
     Amna ajah—ah-hu!
And brings us weather fine and fair,
     Amna ajah, ah-hu!
They continue the whole night through; they run
for several days and nights, till they are so fatigued
and spent that they can no more speak.

This sort of behavior—drunkenness, excessive indulgence
in sensual pleasures—is practiced at particular times in cultures all over the world; compare the American Mardi Gras or
New Year's Eve in many Western cultures. If you want to start
out a fresh cycle of life with a clear conscience, better to purge
all that wicked behavior out of your system before the old
clock runs out. I think the stress preceding all this emotional
relief probably comes from a realization that the winter's food
stock has run dangerously low. What once seemed an abundant stockpile accumulated from the harvest or hunt is now
almost gone. Who knows what the next cycle will bring? What
better time to abandon our woes, release ourselves into free-
dom, and live for today—at least for a while?

Winter comes swiftly in high latitudes. Nights lengthen as
the noonday sun swings lower in the sky, casting longer shad-
ows. Thoughts turn to the hunt and many northerners focus
on chasing bears. Circumpolar constellations act out the chase.
The bear is the quadrilateral bowl of the Big Dipper; his pursuers, their spears held aloft, are the stars of the Dipper's handle.
In another version the entire Dipper outlines the bear's body

Big Dipper Bear Hunt, a Native North American constellation.
(starnameregistry.com, redrawn by Julia Meyerson)

(the bowl) and tail (the stars of the handle), with the fox of
Boötes giving chase.

   To make a story about the hunt memorable, you need to
add realistic details, like autumn leaves turning red during
the hunting season. A member of the Great Lakes Fox tribe
told this bear story to a late nineteenth-century traveler: Once,
when the first snow fell, three men went out on an early morning hunt. They quickly picked up a bear's trail and followed it,
but before they caught up, the bear detected their scent and
darted off to the north. "He's moving toward the noonday
sun," shouted the hunter flanking the bear on the east. The one
who guarded the southern part of the landscape yelled out:
"Now he's running for the 'going-down of the sun.'" Round
and round they went for days, keeping the bear constantly
on the move over the four sides of the landscape, until one of
the hunters happened to look down. There he beheld the surface of the world. They had chased that bear all the way up into
the sky! When they reached the place called "River That Joins

Another," one hunter said to the other two: "Better to go back before it's too late. We're being carried up into the sky." Still they persisted, and by autumn they finally overtook their prey. They killed the bear, skinned it, cut it up, and laid the flesh on boughs of oak and sumac to dry. Since in hunting cultures sharing food is just as important as not stealing it, before dining on any of the flesh, the hunters respectfully slung portions of their quarry into the sky:

> Toward where morning comes they flung the head; they say that just before dawn in winter some stars that came from the head of the bear begin to rise. The four stars in the lead are the bear and the three that follow are the bear chasers. And in the autumn, when the bear lies lowest on the horizon, the leaves of the oaks and sumacs turn red for that is the place where the hunters lay the bleeding body.

N. Scott Momaday, a Kiowa storyteller, recounts a different bear story, one told by his parents when they lived in the shadow of Devils Tower (Wyoming), a massive butte made famous in the 1977 sci-fi film *Close Encounters of the Third Kind*. Eight Kiowa children, seven sisters and their brother, were playing at bear hunting in the Black Hills. The boy pretended to be a bear chasing his sisters through the woods. The girls were pretending to be afraid, so they ran very fast. Suddenly the boy turned into a real bear. Now the girls truly became terrified. They ran faster and faster in fear for their lives. They passed by the huge stump of a tree that spoke to them: "Climb up on me, I will save you." The little girls scampered to the top of the tree stump, which slowly began to rise up into the air. By the time the bear arrived at the base of the stump to kill them, they were far out of reach: "The little girls were borne into the sky, and they became the Big Dipper." If you stand south of Devils Tower (latitude 45 degrees north) and look up, you'll see this myth unforgettably dramatized in the land- and skyscape. Following evening twilight in early autumn, the Dipper is at its low point; it just misses touching the north horizon. The seven stars that comprise it pass just behind the impressive monolith. Concluded the storyteller: "So, you see, we Kiowa have relatives in the sky."

The Bear-Dipper link is surprisingly widespread in the Northern Hemisphere. There is even an ancient Greek version, though it is far different in context from other bear stories. The story opens with Callisto, queen of Arcadia. Known for her love of hunting, she often accompanied Artemis, the goddess of hunting (and chastity). One day while out on the hunt, Callisto was ravished by Zeus. Fearing reprisal for losing her virginity, she kept the encounter a secret from her friend for as long as she could, until she became heavy with child. When Artemis found out about the affair, she punished Callisto by taking away her feminine form and changing her into a bear. Her life dramatically altered, Callisto wandered the forest as a wild animal until she was captured by hunters and brought to their king as a gift. One day she entered one of the temples dedicated to Zeus. The gift-givers pursued and attempted to slay her for illegally trespassing. But Zeus took pity on her and placed her in the sky. He named her Arktos (bear in Greek), whence our word arctic. (In an alternative ending to the story, Callisto's son, while hunting in the woods, was recognized by his transformed mother. When, overcome with maternal joy, she charged at him, he shot her.)

Celestial bear stories recorded across the high and middle northern latitudes migrated south. This is likely the ancient source of the Greek bear constellation. Siberian hunters may even have followed the Dipper as it moved nightly in the sky when they crossed (in the same direction) the Bering land bridge that once linked Asia with the Americas. When the migration to the Americas began thirty thousand years ago, the Dipper was in approximately the same location relative to the pole as it is today.

Outsiders to tales about polar constellations can have strong reactions, dismissing them as wrongheaded or too inaccessible to study. Given the imaginative star stories we've encountered so far, such a dismissal seems to be missing a larger point about perspective.

Thinking analytically, when we look at the stars circulating perpetually around the fixed pole, we attribute the cause of their motion to be a reflection of our collective understanding that the earth rotates: the stars move in relation to us in the sky the same way that houses, trees, and hills flanking the highway seem to move backward as we zoom forward. We can prove this effect scientifically by experimentation.

There is no such cause and effect explanation involved in the *associative* sort of thinking that underlies the stories I've been narrating. Associative thinking likely originated in prehistoric times as a way of recalling perceived patterns in the universe, and as a means of fitting patterns and events into a scheme that would cover all the mutual connections or influences that might occur among its parts. Basically, it involves list-making. For example, just as I might list the parts of my body from head to foot, so too might I organize the stages of my life from birth to old age, by passing along the continuum from top to bottom and by identifying each stage in one list with a segment or joint in the other. Thus, I could say that my head represents childhood and my foot old age. By listing all the events that make up my universe of existence—say, the elements, the seasons, the constellations of the zodiac—I can fashion a hierarchically ordered system whose core is based on a principle of association. Autumn leaves turning red in the story of the bear hunt fit into the list of events that happen in that season on earth as well as a list of what takes place in the sky, which we might otherwise regard as separate domains of human experience. For early human communities, connections like these offered reassurance, predictability, and advantages for survival, because they helped people identify and remember the turning of the seasons.

7

Star Patterns in the Tropics

Have you ever noticed that Scorpius's tail looks like a hook, especially when it rises in the southeast? Hawaiians call it Maui's Fishhook. Maui was a mythical hero, a fisherman who loved casting his line in the coral reefs below Mount Haleakalā, which is all that once existed of the islands. But Maui was never very good at securing a catch, and his brothers would often tease him about his failures. What they didn't know is that Maui possessed a magic fishhook, though he kept it a secret and saved it for important uses.

One day Maui decided to play a trick on his brothers. He deliberately caught his magic hook on the ocean bottom. "Paddle as hard as you can," he told them. "Looks like I've got a huge fish." As they did, Maui hauled up a large island, the island of Maui. His brothers were so busy paddling they didn't even notice. So Maui repeated the trick and pulled up another, even bigger island—Hawaii. Then another, Oahu; then Kauai; then Lanai; then Molokai, Niihau, Kahoolawe, and Nihoa. And that's where the Hawaiian Islands came from.

Another version of the story continues to a different ending. In it, Maui does his fishing off the coast of the big island, and he orders his brothers not to look back, lest the expedition fail. When a bailing gourd appears on the water, Maui instinctively reaches out for it and places it next to him in the boat. Suddenly a beautiful water goddess materializes. His brothers can't resist turning around to gaze at her, and as they do, the fishing line goes slack and all the islands Maui's brothers have hauled up from the deep sea sink back partway. That's why Hawaii is a chain of islands instead of a single large land mass.

One day Maui's mother complained that the days were too short: there wasn't enough daylight to dry her clothes. "Why is the sun moving so fast?" Maui set out to solve the problem by capturing the sun. When he did, however, the sun begged to be let off the hook, promising to slow down and make the summer days longer. You can see for yourself that the promise is still kept. There's Maui's hook adjacent to the spot in the zodiac where the sun passes in predawn midwinter skies, portending longer days. That's the place where the sun gets hooked every year.

There's a big difference between Maui's Hawaiian sky and that of the Arctic dwellers we encountered in the previous chapter. To see it, head south from Inuit territory, where the Pole Star stands 70 degrees high in the sky and stars gradually lift off the ice fields in the east and set at equally low angles in the west. Halfway to the equator, at latitude 45 degrees north (where southern Europe, the northern United States, and northern China lie), star trails make 45-degree angles with the horizon, and Polaris lies halfway between the skyline and the overhead point. By the time you reach the Tropic of Cancer at 23.5 degrees (northern Africa, Mexico, and India), Polaris is much closer to the horizon (23.5 degrees) than it is to the overhead point (66.5 degrees). Star trails become steeper as you get closer

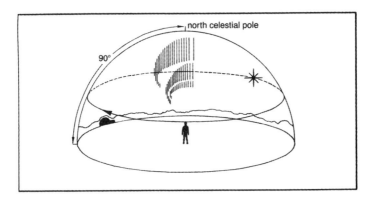

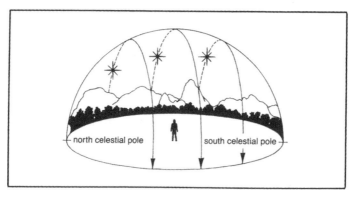

Comparing star trails in polar (*top*) and tropical (*bottom*) skies.
(Peter S. Dunham)

to the equator until, at zero latitude, the Pole Star becomes virtually invisible at the horizon and sky objects rise and set along vertical lines. This straight up, over, and down motion, with you at the center, stands in stark contrast to the round and round circular movement about a fixed point on the sky witnessed by those who live in the middle and higher latitudes.

Now imagine you're gliding eastward in a canoe along a vast tropical ocean. You're at eye level with a featureless horizon, no sighting posts or pathways to guide you along your course. You watch the stars ascend vertically in front of you and plunge straight down into the ocean behind you. You might begin to wonder: am I moving, or are the sky and ocean slipping by me as I remain stationary? Polynesian navigators claim they experience the sky and ocean as slipping by. An unnamed early nineteenth-century missionary copied down this fragment of a Tahitian sea chant that describes the way land and sky imagery are lifted up out of the horizon in a moving landscape:

> The sea casts up Basket Island alone, and Angular Island alone. Just over the sea stands Aldebaran, weeping for god Rio.
>
> Bear thou on! And swim where? Swim toward the declining sun, swim toward Orion.

Being a navigator in the western Pacific islands of Oceania in the eighteenth and nineteenth centuries was as respectable as being a neurosurgeon today. An eighteenth-century European sea captain who visited the islands remarked that geography, navigation, and astronomy were known only to a few. Getting from one place to another in an environment made up of 99 percent water, on an ocean affected by shifting winds, swells, and currents, required carefully cultivated skills. Being familiar with the sky was a major component of those skills. In eastern Micronesia, such an expert was the *tiborau*, a perspicacious skywatcher and an experienced navigator who could locate and memorize the shapes and positions of linear star-to-star constellations on an imaginary three-dimensional compass. This person knew how to identify a chain of stars appearing over the same point on the horizon throughout the night and use that knowledge to steer his boat to a particular island. How exactly did the tiborau master these skills? And why these unusual linear constellations?

Arorae is a small Pacific atoll in the Gilbert Islands. Its northern shore is dotted with half a dozen pairs of rough-cut slabs, each about the size of a person, laid out in parallel along the ground. One pair points to the neighboring island of Tamana fifty miles distant, another to Beru Island eighty-five miles away, and a third to Banaba, 440 miles over the horizon. Islanders call these slabs "Stone Canoes" or "Stones for Voyaging," and they were used to set directions for inter-island navigation, as well as for instruction. A modern navigator who once visited a tiborau was shown a functioning model of a stone canoe located behind the family residence. Built by his father and based on an earlier version constructed by his grandfather, the structure measured about five feet east to west by four feet north to south. It was equipped with a rectangular stone seat in the center and triangular rocks of different sizes and orientations representing the magnitudes and directions of ocean swells. A chunk of brain coral astride the seat symbolized the god of the sea. At the time (the mid-1960s), the tiborau still employed this simulator to teach his daughter how to navigate. Unfortunately many of the indigenous star names provided by island informants were never taken seriously enough by outsiders to have been transcribed onto Western star charts, so we don't know the identity of most of the stars in their system.

The parallel slabs that make up a Stone Canoe align with the place where certain stars will appear or disappear on the sea horizon at different times of the night. For example, at sunset in August the bright star Regulus lines up with the Tamana stone, while, at midnight, Arcturus gives the same bearing. The

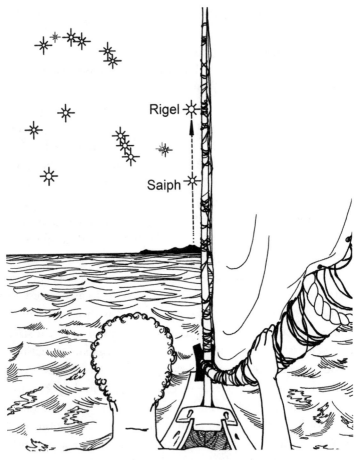

A Polynesian navigator guided by a linear constellation.
(Drawing by Julia Meyerson)

navigator simply memorizes a *havienga,* or star path, consisting of a long chain of stars associated with the island he wishes
to visit, then steers the canoe toward the appropriate linear
constellation. The star rise and set positions become the points

on a kind of three-dimensional memory compass developed through oral tradition and trial-and-error observations. Investigators have recognized three-, five-, seven-, and nine-path navigational star systems in use over three thousand miles of the Pacific Ocean from the Trobriand Islands, off the eastern coast of New Guinea, to Samoa, which lies halfway between Hawaii and New Zealand.

Skilled navigators in the Caroline Islands, in the southwest Pacific, imagined the sea horizon to be divided into thirty-two points, each opposite another that is also connected to the observer, who is located at the center. For example, rising Vega (northeast) is the opposite of setting Antares (southwest); Aldebaran rising (east-northeast) opposes the place where Orion's Belt sets in the west-southwest; and the Southern Cross, which rises in the south-southeast, has its reciprocal in the pointer star Alpha in Ursa Major (north-northwest), and so on.

Hawaiian navigators constructed model star compasses out of hollowed-out gourds, instructions for which appear in an 1865 guidebook:

> Take the lower part of a [rounded-out] gourd . . . on which several lines are to be marked . . . These lines are called "Na alanui o na hoku hookele" (the highways of the navigation stars), which stars are also called "Na hoku ai aina" (the stars which rule the land). Stars lying outside of these three lines are called "Na hoku a ka lewa," i.e., foreign, strange, or outside stars. The first line is drawn from "Hoku paa" (North Star), to the most southerly of "Newe" (Southern Cross). The portion to the right or east of this line is called "Ke alaula a Kane" (the dawning, or the bright road of Kane), and that to the left or

west is called "Ke alanui maaweula a Kanaloa" (the
much traveled highway of Kanaloa).

Other markings on the gourd's edge include the key navigation stars.

Navigating in this way must have required generations of
experience. To solve the problem of translating star positions
into directions for long-distance sailing, maritime astronomers
of Oceania developed two concepts that possess no analog
in the cosmology and astronomy of other civilizations: linear
constellations and an imaginary star compass. Both these creations of the human mind seem to have been prompted by the
environment. The orientation of the sky in the tropics made it
possible to navigate using techniques that circumvent the con-
ventional magnetic compass and other astronomical contriv-
ances of contemporary culture. Think of the problems a North
Pacific or South Atlantic sailor would have setting sail by linear
constellations. As soon as a guide star appeared, it would begin
to move laterally relative to where it first appeared on the horizon. Unless a substitute star was immediately available, the
navigator's course would deviate radically from a straight line.
For example, if one attempted to use a series of "linear" guide
stars to sail from New England to Great Britain, the ship would
be thrown off course in short order.

Practical-minded Oceanic people took a fact of geography and turned it to their advantage. They used horizon-based
astronomy because at equatorial latitudes this system worked
well. Part of the system involved using the overhead point. The
navigator would memorize *fanakenga* stars, those that passed
the overhead point, one each for a particular destined island.
(In modern astronomical parlance, the declination, or the angular distance north or south of the celestial equator of a star

that passes overhead, turns out to be the same as the latitude of the observer.) Put in contemporary terms, a navigator would know the latitude of a particular destination had been reached when the arc of the guide star's course crossed the overhead position. Thus the Tongans say that Sirius is the fanakenga of the Fiji Islands (latitude 17 degrees south), and Altair is the fanakenga of the Caroline Islands (latitude 9 degrees north). Based on modern trial and error, navigators have discovered that under calm conditions, you can estimate geographic latitude to within one-half a degree (about thirty-five miles) by sighting fanakenga stars.

As late as 1890, an unnamed Tahitian navigator recited these instructions to a visitor for the first time:

> If you sail for Kahiki [Hawaii], you will discern new constellations and strange stars over the deep ocean. When you arrive at the Piko-o-Wakea [the equator], you will lose sight of Hokupua [the Pole Star] and then New [?] will be the guiding star and the constellation of Humu [?] will stand as a guide above you.

Unfortunately, the English names of the celestial guides remain lost.

As far as we know, longitude (like latitude) was not conceptualized by indigenous sailors of the mid-Pacific. There simply was no method known for dealing with longitude, and, more importantly, no motivation to create such a method. Although Old World navigators found it necessary to use time-keeping mechanisms, ranging from sand clocks to (later) chronometers, in order to chart the east-west segments of a course, the people of Oceania employed their knowledge of wind and ocean currents combined with astronomical observations; that is, they relied on purely natural forces to hone navigation into a fine art. That they managed to get about successfully proves that knowledge of longitude and latitude—so familiar in our culture—is not essential for all skilled navigators. Deep-sea voyaging across the southeast Asian archipelago without techno-aids may seem very difficult, but modern anthropologists and adventurers who have used native technical skills to reconstruct and carry out such voyages themselves leave no doubt that it could have been done.

Explaining these navigational concepts to inquisitive outsiders, however, has always proven difficult. In a classic example of failure to communicate, a German sea captain became frustrated over his inability to understand a tiborau's explanation of how his navigation system worked. The captain wrote in his 1897 log that the tiborau

> once told me with seeming frankness that I was the dumbest churl he had ever seen; daily he told me the same, and that every day I came again with the same stupid questions; generally he would have no more to say to me, and only a glass of [sherry], which the old man loved, would make him friendly again.

A question often raised about the cultures of Oceania is how they first managed to get to their distant islands from the mainland, since in many instances, hundreds of miles separate segments of the great island chain. Yet there would have been no shortage of practical motives beyond pure adventure for conducting long-distance trips across vast expanses of water: for instance, to trade material goods, conduct raids or warfare, extend the reach of a chiefdom, or obtain food and other items from uninhabited islands. One group of Cook Islanders was known to have traveled over 185 miles just to procure supplies of exotic birds' eggs. Economic disruption at any time on the mainland would have offered a reason to move out into the more sparsely inhabited or totally uninhabited islands far off the coast.

Indeed, given how essential water-based travel was for the community's survival, celestial navigation by linear constellations may have become the skill that the civilized peoples of Oceania treasured most. Today, by contrast, we travel by car or we entrust our itineraries to bus drivers, train engineers, plane pilots, or the map on our iPhones. Thanks to technology, none of our long-distance travels require us to look at the sky. Only the handful who navigate for pleasure without the use of technology can truly appreciate how important sky knowledge once was for setting direction.

A story from Polynesia, best told under winter skies, demonstrates the imaginative, sea-themed mythological imagery that is distinctive among tropical cultures (while echoing some of the Greek moral lessons about boastful Orion). There people once called the Pleiades Matariki, or "Little Eyes." They say that once Matariki was a single star, brightest of all in the sky—so bright that when it rose its reflection sparkled on the sea and dazzled everyone. Matariki became very conceited about his brilliance and bragged to the other stars, saying, "I am more brilliant than all of you, even more brilliant than all of the gods themselves." Tāne, the one who guards the four pillars that hold up the heavens, became angry and decided to drive the boastful one to the dark regions. He sought out the help of two neighboring stars: Mere (Sirius), who was the second brightest star in the sky and had little sympathy for his rival, and Aumea (Aldebaran), who resided so close to the brilliant star that he had become annoyed and embarrassed over constantly being outshone.

One night the trio sneaked up behind their victim. But Matariki spotted them and took off, hiding in the waters of the nearby Milky Way river. Mere climbed up to its headwaters and diverted the water's course, leaving the fugitive unprotected. Again Matariki fled, passing beyond the arches of heaven as he began to outdistance his pursuers. In desperation, Tāne grabbed hold of Aumea and hurled him against the luminous one with such a powerful force that when he hit his target it broke into six pieces. When the pursuers left the scene, the remains of the once brilliant star limped back to where they belonged. Mere became brightest in the sky and Aumea is no longer dimmed by a nearby rival. As for the Little Eyes, some say that now they whisper among themselves that they are even more noteworthy as six than they were as one. Perhaps when the wind dies down and the seas are calm, they lean down close to the surface, view their images in the water as they are now, and vainly convince themselves that they still have no equal.

8

Empire in the Sky

An old Pawnee story narrated during the First Thunder Ceremony tells of Fools-the-Wolves, the coyote deity who was the grandfather of the Morning Star, and Paruxti, grandfather of lightning, controller of all the fires and lighter of the stars that enabled people to go out at night. Paruxti had his own star, the Evening Star, that rivaled Fools-the-Wolves' Morning Star. Fools-the-Wolves became jealous of Evening Star's luminescence and power. They say that originally Paruxti placed all the constellations on the ground, intending them to live there forever as an immortal race. But Fools-the-Wolves sent a pack of wolves to steal Paruxti's lightning bag. The people killed the wolves, bringing death into the world for the first time.

The people decided that they wanted their dead to return after only a short time away. But when some returned and others decided to stay away, Fools-the-Wolves set out to make them all remain dead. He built a medicine lodge where all the souls were received in a whirlwind. Once they were inside, he caused the wind to shut the door, and ever since all people must die. To combat his bad reputation for bringing death into

118   Empire in the Sky

existence, they say that the coyote deity's star enlists the help of the Snake Star (Antares), who sends snakes to earth to kill those who tell bad stories about him.

The father of the Lakota keeper of the original sacred pipe once told a visitor: "[Earth and sky] are the same because what is on earth is in the stars, and what is in the stars is on the earth." (Modern cosmologists would agree—everything in the universe is made of the same stuff!) The Lakota used the "As above, so below" astrological slogan to explain their idea of mirroring: the earth map consists of hills, valleys, rocks, and crops, and the skyscape constellations of their sacred stories connect to specific places like these in the landscape. Their stars are situated in or close to the zodiac, the idea being to synchronize the Lakotas' arrival at various sites to coincide with the appearance of the sun in the corresponding constellations, or, as they say, to follow the sun's path on earth. For example, Devils Tower is connected to Mato Tipila (Castor, Pollux, and six surrounding stars). And the direction of the winter camps to the south of the tower is marked out by a pair of constellations made up of what Western viewers know as Triangulum and Aries, also to the south.

Think of what might have happened had English settlers never come to the New World, had the Lakota not been uprooted, had they consolidated their influence over surrounding tribes—indeed had they built a great city to reinforce their power and influence. Maybe the plan of that city would have retained, and perhaps even elaborated on, many of the cosmic hallmarks that governed the sacred spaces of the village. This never occurred on the Great Plains of North America, but it did happen elsewhere in the world, even in the humble house.

People who live on Kiribati, one of the Gilbert Islands in the central Pacific that straddle the equator, say that the sky is their house. They segment the sky house (*uma ni borau*) into named zones formed by slicing the sky dome several times in the east-west vertical direction and segmenting it with another set of lines running parallel to the horizon. They call the vertical lines ridge poles (the north-south meridian passing overhead) and rafters (small circles), and they liken the horizontal arcs to crossbeams, or purlins. They locate a star by its position in one of a collection of imaginary architectural boxes that partition the cosmic domicile.

Kiribatians use their sky-house analogy not just to locate stars, but also as a seasonal clock. For example, when the Pleiades reach the first purlin in the east about an hour before sunrise, they know that the sun lies at the June solstice, which times the beginning of the wet season. A similar sighting of Antares signals the arrival of the sun at the spring equinox. As one elder told nineteenth-century anthropologist Sir Arthur Grimble: "When you see Rimwimata (Antares) in the middle, between the ridge-pole and the first purlin to westward, you know that the sun is on his Bike ni Kaitara (Islet of Making-face-to-face)," the Kiribatian term for the vernal equinox, which commences the dry season.

The Navajo forge similar architectural connections between the sky and the place where they live. They call it the *hogan*, after *ho* (place) and *ghan* (home). The Hogan of Creation was made by Black God, whom we met in the Navajo story of the Pleiades. He populated it with stars and it serves as the archetype of all the hogans that we, their descendants, create here on earth. Out of reverence to Black God, all hogans need to be oriented in a direction sympathetic with the movement of the stars across the night sky. The roof of the hogan must be domed like the sky, and it must be round like the sun (*haʼaʼaah*, meaning "the round object that moves in regular fashion"), the source of light and heat. Above all, the hogan must face east, the place where the sun begins its journey every day.

Every hogan has four posts, one positioned in each of the cardinal directions, just like each of the four mountains that hold up the sky. The walls of the hogan are vertical, just like the mountains. When you enter the hogan you must always move clockwise, imitating the way the sun moves. The interior is not physically divided, though four specialized areas, or recesses—north, east, south, and west—are recognized. There is the place where you cook and eat, in the east; where you entertain daytime visitors, in the west. The place where you pray and show reverence lies in the north; the place where you work, in the south. The fifth direction, the center, symbolizes the sky that surrounds the central fireplace, the sun.

Navajo chanters connect the duties of the household to particular constellations. They interpret the Big Dipper as the male who revolves and Cassiopeia as his female counterpart. Forever they turn around Polaris, the central fire, the igniter, in the sky hearth. They are the old married couple who set a positive moral example by staying at home and tending the fire. They remain always with their families and carry out their responsibilities to the group.

Like the Kiribati and Navajo houses, the Skidi Pawnee lodge, which can hold an extended family of up to thirty residents, mimics heaven's architectural plan. Its domed shape imitates the sky, down to the precise arrangement and orientation of its components, with the doorway on the east, four directional support poles, and a carefully positioned smoke hole through which inhabitants can watch the stars. In his book on

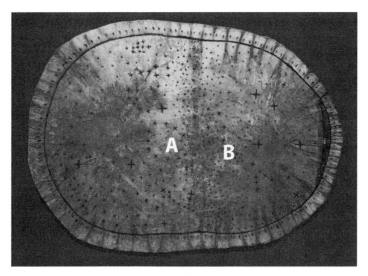

A nineteenth-century Skidi Pawnee star map.
Note the central positions of this culture's two most
important star groups, (A) Corona Borealis and (B) the Pleiades.
(Werner Forman/Universal Images Group/Getty Images)

life among the Pawnee, published at the turn of the nineteenth
century, ethnologist James Murie wrote: "At a certain hour, a
priest . . . looked up through the smoke hole. If he could see
[certain] stars directly above, it was time for the planting ceremonies." Unfortunately Murie, born of a Pawnee mother, gave
few details; but we do know from later informants that the Pleiades, a Pawnee symbol of unity, would be first glimpsed briefly
just before sunrise in late July through a lodge smoke hole by
an observer positioned at the wall, along the lodge's axis of
symmetry. They would then be seen again just after sunset
around the time of the winter solstice. The Chiefs in Council
(our constellation of Corona Borealis) shone through the smoke
hole in direct opposition to the Pleiades' appearance. This may

explain their prominent location in space on native star maps. The Pawnee lodge serves both as shelter and astronomical observatory or planetarium—nature's living classroom. Within its warm confines, children could view for themselves how the celestial scenery was used to dramatize the real-life stories and moral tales they were told.

The Kiribati, Navajo, and Pawnee domestic architectures are three examples among many that demonstrate the close parallels between the two roofs that lie over our heads. Think of how different are modern ideas of what makes an ideal home. To judge from its architectural plan, today's domicile is less communally arranged; for example, each family member almost always has a private space in the household. Few actually live in the living room or dine in the dining room. Environmental factors, too, barely play a role in the placement and orientation of the houses we live in. Water can be piped in from afar, trees of all kinds can be planted, and a mountain or ocean view costs extra. There may even be a true sunporch, but where are the stars in relation to our living space? For most modern people, this doesn't really matter.

The Skidi Pawnee and Lakota sought a grander connection to the cosmos when they designed their homes and community buildings; in fact, they patterned entire villages and landmarks after the positions of the constellations. One of their myths tells of the stars originally being brought to earth, where they established the locations of their camps according to their stations in the sky. The Skidi assigned each village a patron star and made the villagers responsible for conducting rites tied to it. For example, ceremonies at the shrine of the village farthest to the west, under the rule of Tirawa, god of life and knowledge, were the first in the seasonal calendar to be performed. The initial rite, First Thunder Ceremony, happened when springtime thunder was first heard, around the time of the equinox. The four bowl stars of the Big Dipper, which pivot about the North Star, mapped out the positions of the five central villages on the ground relative to the center. Their shrines represented the affairs of the people. Each village participated by supervising a ritual connected with tribal matters, such as planting, harvesting, hunting, installing leaders, and conferring honors on warriors.

As the order of the seasons progressed, the rites shifted from shrine to shrine, from west to east. Under the Morning Star, the eastern village came last: its job was to conduct a sacrifice that connected the world above with the world below to ensure the perpetuity and productivity of all life on earth. Then the cycle would repeat itself. We don't know much about what went on during these ceremonies. Pawnee leaders have always been fairly secretive about how they conduct their sacred rituals. One elder, however, did indicate that they "give an account of creation, the establishment of the family, and the inauguration of rites by which man would be reminded of his dependence on Tiráwa, of whom he must ask food."

Chinese dynastic society was highly bureaucratic and special care was taken to document events in the night sky. Royal family histories are filled with lengthy chapters on astronomy that include details about where and when celestial objects appeared or disappeared—their color, brightness, direction of motion—and especially where and when they gathered in the same place. These histories forge deep connections between celestial and family affairs; thus one Chinese historian and court astrologer explained that when the "heavenly minions" (the planets) gather, either there is great fortune or there is great calamity. He knows this, he writes, because when they once gathered in Roon (Scorpius), the Zhou dynasty flourished (first millennium BCE); but when they gathered in the Winnowing Basket (Sagittarius), Li Qi, who was known for his reign of decadence and deterioration, became emperor (in 314 CE).

As we learned earlier, the Western world acquired its constellation names from Greek legends, and although some of the original forty-eight constellations, such as Andromeda and Perseus, Orion and Scorpius, are mythologically linked, most stand alone. The Chinese system is quite different. When the Chinese turned their eyes to the north they saw nothing like a pair of wheeling bears to remind them of the impending hunt: instead, a single grand scheme connected all 283 of their patterns in the sky. Like the heavenly houses and lodges of the Navajo, Pawnee, and Lakota, the Chinese sky was designed as a celestial empire, imagined to reflect the most important house in imperial China. Most essential of all were the three Enclosures of the Northern Sky. They are centered on the emperor, who is represented by the immovable Pole Star.

Confucius compared the emperor's rule to the Pole Star, noting that just as the emperor was the axis of the earthly state, so his celestial pivot was the polar constellation. The economy revolved around the fixed emperor the way the stars turn about the immovable pole. According to one legend, the Divine King was born out of the light that shone onto his mother by the Pole Star. Clearly, the fixed location of the polar axis became a cosmic metaphor for the constant power of the state. Like the emperor's domain on earth, the representatives and other details of his imperial court, the so-called Purple Palace, consisted of the stars closest to the pivot points of all motion, those stars that never set. Each of these celestial functionaries had a terrestrial counterpart.

Four of the seven stars in what we call the Little Dipper,

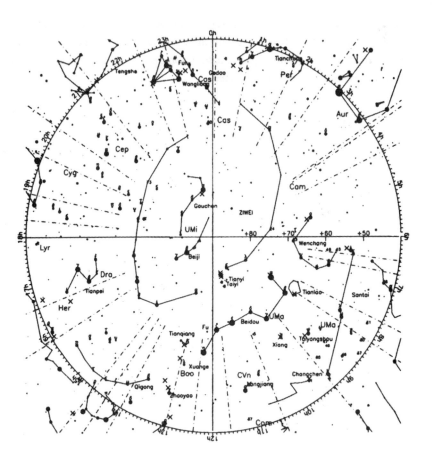

Constellations of the Chinese Purple Palace. The emperor,
cast as the immovable Pole Star, is the brightest star in the
Gouchen group. (From S. Xiaochun and J. Kistemaker,
The Chinese Sky During the Han [Leiden: Brill, 1997], 70)

plus two other stars, constituted Gouchen, the "angular ar-
ranger" of the second-century BCE Qin dynasty. One member
of the group was the crown prince, who governed the moon,
while another, the great emperor, ruled the sun. A third, the
son of the imperial concubine, governed the five planets, while
a fourth was the empress, and a fifth the Heavenly Palace itself.
When the emperor's star lost its brightness, his earthly counterpart would ease up on his authority, while the crown prince
would become anxious when his star appeared dim, especially
when it lay to the right of the emperor.

The four surrounding stars of the palace proper are Beiji,
the Four Supporters. On Chinese star maps they appear well
situated to perform their task, which is to issue orders to the
rest of the state. The Golden Canopy is made up of protective-
looking chains of stars corresponding to the pole-centered
stars of our constellations Draco, Lynx, and Camelopardalis.
Beyond the palatial inhabitants and emissaries lay the stars of
the Northern Dipper, Beidou. More concerned with realizing
celestial principles in the earthly realm, these Seven Regulators
are aptly situated to descend close to earth so they can inspect
the four quarters of the empire.

During the time of the Five Dynasties (tenth century CE),
the Big Dipper was conceived as the administrative center. It
housed the famous Daoist priest and deity Zhang Xian, known
in his previous incarnation as Emperor Meng Chang, noted
for his extravagant (and immoral) way of life. His chamber pot
was made of gold and studded with pearls. When Zhao, the
first Song dynasty emperor, defeated him and occupied his
palace, he beheld the glittering chamber pot and announced
that anyone indulging in such extremes deserved to be shamed.
Accordingly Zhao imprisoned Meng Chang and his concubine, Lady Pistil. After he caused further trouble for his cap-

tors, the vanquished king was ultimately assassinated. In memory of her lost lover, Lady Pistil painted a beautiful portrait of him. When Emperor Zhao visited her, he asked, "Who is the subject of this portrait?" Fearing reprisal, Lady Pistil answered, "It's a picture of Zhang Xian," a popular deity of the Sichuan province who would grant a son to those who made a sacrifice to him. Thus, the effigy of terrible Meng Chang came to represent an honorable god.

Another version portrays the Big Dipper as the carriage of the Great Theocrat who periodically wheels around the central palace to check things out. Its stars are the source of yin and yang, the twofold way of knowing that resolves the tension between opposing polarities: male and female, light and dark, active and passive. Yin and yang wax and wane cyclically with cosmic time. Together they make up the potentiality of the human condition. For every affair of state, the starry winds of good and bad fortune blow across the sky.

Other parts of the Purple Palace, some represented by faint stars, include the Royal Archives Official, Female Protocol, Maids-in-Waiting, Chief Judge, Three Top Instructors, Eunuch, Prime Ministers, Celestial Guest House, Celestial Kitchen, and the Celestial Bed. No earthly concerns of life go unaddressed in the Chinese empire in the sky: there's a Star of Betrothal, a Mental Illness Star, a Rheumatism Star, a Star of Pleasure, a Lawsuit Star, a Prion Star (the source of itching), and a star called the Orphan Star, which enables a woman to turn into a man.

There is more than one Dipper in the Chinese sky. A second one is located far from the pole, in our southern constellation of Sagittarius. They say that this Dipper plays chess with the Dipper in the north, with the outcome of the game being a matter of life and death. It all started when the Daoist priest

Guan Lu met Yan Chao, a young man. He sensed that, though
Yan Chao looked well, he would soon die; so he decided to help
him by offering the following advice:

> Go home right away and prepare a dinner of veni-
> son and a bottle of wine. Take it to the biggest tree
> in the southern wheat fields. There you'll find two
> old men playing chess. Serve them and be silent.
> They will help you.

Yan Chao did exactly as he was told. But the two old men
seemed so absorbed in the game that they paid him no attention. Though they ate, they didn't seem to enjoy the feast. After
a long while, one of the chess players noticed the young man
standing there and casually asked, "Who are you and what are
you doing here?" Yan Chao told his tale and begged them to
save him from impending death. "Let's check your record of
death," responded the other chess player. When the old man
saw "nineteen years" as the projected age at death, he crossed
it out and wrote in the margin "ninety years." Thrilled with the
result, Yan Chao went happily on his way. When he returned
home, his friend Guan Lu explained:

> The old man sitting on the north is the Northern
> Dipper, the one to the south is the Southern Dip-
> per. Southern Dipper is responsible for birth, and
> Northern Dipper is responsible for death.

To harmonize the arrangement of the royal capital on
earth with the local contours of cosmic energy, the king would
call in to perform the art of feng shui a geomancer, an interpreter of markings on the ground or patterns formed when

handfuls of dirt are cast into the air. This expert would decide where to place each building site and how to arrange everything in and around it. His sources of cosmic knowledge included the local magnetic field, the paths of streams, and the land forms; he might also consult oracle bones, engraved pieces of ancient bone and shell used in divination. Sometimes workers would be required to transform the landscape, removing vast numbers of boulders and planting forests of trees to regulate the disposition of yin and yang energies passing in and out of the site. (You can still see such architectural modifications in the skyscrapers of Shanghai and Beijing.) Acquiring proper urban form depended on getting things right with the natural environment—especially the cardinal axes. To function properly, the city needed to be aligned exactly north-south and arranged in the shape of a perfect square. According to historian Paul Wheatley, "They erect a post at the center, taking the plumb bob lines to ensure its verticality, and with it observe the sun's shadow. [Then] in the nighttime they study the pole star so that true east and west is precisely fixed." As Chinese historian Joseph Needham concluded, this was the cosmic pattern that responded to the harmonious cooperation of all beings, a pattern "determined not by some superior authority but rather by the internal dictates of their own lives."

Beijing is still oriented according to its ancient cosmic plan. If you stand in Tiananmen Square you can line up the Bell and Drum Towers, the Monument to the People's Heroes, even the Mausoleum of Mao Zedong, on a perfect south-north axis. Continue that line and you'll discover that it runs through the gates of the old city. Today the cosmic axis is defined by a marble walkway that marks the Imperial Meridian, or north-south line. The Hall of Supreme Harmony, which houses the emperor's throne, lies at its northern terminus; there the line of sight lifts straight up from the place where the earth meets the sky to the circumpolar region.

The cosmic arrangement of Beijing has a parallel in a well-known Christian prayer, "on Earth as it is in Heaven," which is yet another variant of "as above, so below." Indeed, we need go no farther than the capital of the United States to find a shared communal ideology reminiscent of the Chinese city in the sky. A 1792 essay in a popular magazine described the recent laying out of Washington, DC: "Mr. Ellicott drew a true meridional line, by celestial observation, which passes through the area intended for the Capitol. This line he crossed by another, running due east and west, which passes through the same area."

The technique of drawing a line from the median position of Polaris perpendicular to the sun's daily path and extending it to the south is exactly the method used to construct ancient Beijing. Washington is the quintessential planned American city, a product of the eighteenth-century French Enlightenment. Conceived by French-American military engineer Pierre Charles L'Enfant, its grandiose design was intended to rival the great cities of Europe. Washington was originally laid out in the form of a perfect square ten miles on a side, with its points aligned to the cardinal directions. The Capitol replaces the pyramid or ziggurat that once stood at the center of ancient cities, and the cathedral, temple, or mosque that marks the center of later ones.

Whether monarchy or democracy, every government must establish its connection with the gods, wrote the Enlightenment philosopher Jean-Jacques Rousseau. And Washington's spoked wheel and overlying rectangular grid evokes in the pilgrim—today's tourist—a sense of the national mythology. The city's processional along Pennsylvania Avenue from the Capitol to the White House and its inviting line of sight from the Washington Monument along the Mall to the Capitol mimic the ancient custom of laying out ceremonial ways and pilgrimage routes. Washington presents itself as a center of power, a junction of sacred and secular spaces that imitates the geometry of the universe and the harmony of the worlds. As ancient Beijing was to the Chinese, Washington is every American's city in the sky.

9

Star Ceilings and
Mega-Constellations

Cetus, the sea monster, is the biggest constellation in the sky and a central character in a sky story that responds to the question, Who's the fairest one of all? It begins with Aethiopian Queen Cassiopeia's boast that her daughter, Andromeda, an ordinary mortal, is more beautiful than the divine Nereids, sea nymphs and daughters of the sea deity Nereus, who is a friend of Poseidon, the sea god. Her bragging escalates, prompting Poseidon to send Cetus to ravage the kingdom's shores and devour the child.

Crossing the Mediterranean Sea in ancient times was fraught with danger. Reports of extraordinary shipwrecks left coastal dwellers imagining its waters crawling with all manner of hideous creatures. Recall the multiheaded winged serpent Leviathan of the Hebrew Bible killed by Yahweh, who saves the people from starvation by offering them the flesh of the sea monster. And Scylla, who patrolled one side of the dangerous narrow strait (opposite the whirlpool Charybdis), between the Italian peninsula and the island of Sicily. A beautiful maiden transformed into a four-eyed, twelve-tentacled, six-headed, shark-toothed beast who sank ships and devoured their crews, Scylla has come to represent one of two places between which you'd best not find yourself.

Such stories gave the rulers of the African kingdom good reason to be terrified. They knew Cetus to be a gigantic sea monster with the head of a whale, razor-sharp teeth, and the coiled tail of a serpent. Poseidon's action drove Cassiopeia and her husband to consult the trusted oracle of Apollo for advice. This is not a threat to be taken lightly, advised Apollo. You will find no resolution unless you sacrifice your daughter to the monster, as compensation for your vain behavior. So the parents stripped their child naked and chained her to a rock off the coast of Joppa (Jaffa today, where they say you can still see the site).

Meanwhile, the warrior Perseus was on his way back from ridding the area of another threat, the gorgon Medusa—a winged female monster with a hairdo of venomous snakes, whose mere gaze could turn mortals into stone. Perseus slew the monster by cleverly sidling up next to her while tracking her reflected image in his highly polished shield. Afterward, Medusa's head in hand, he happened upon the young princess in her perilous state, on the verge of being devoured by Cetus. In the nick of time Perseus plunged his sword deep into the heart of the hideous sea monster and set Andromeda free. They say he later married her and the couple raised nine children (seven sons and two daughters), whose descendants ruled Mycenae. To offer a suitable stage for telling the story, the gods placed Andromeda alongside Perseus in the northern sky, not far from Queen Cassiopeia and King Cepheus. But they positioned Cetus the sea monster safely to the south, among other watery-sounding constellations like Pisces the fish, Eridanus the river, and Aquarius the water bearer.

Cetus was my favorite constellation. When I was ten years old I couldn't get enough stargazing, so I cut star shapes out of colored cardboard and glued them to my bedroom ceiling, much to my mother's chagrin (we lived in a rental). I bequeathed center stage over my bed to Cetus, with Orion and its bright blue Rigel on my east wall, and Scorpio on the west. All night long I slept securely enveloped in stars.

Today, if you're out of touch with the real stars, you can content yourself with what earthly manufacturers have to offer by purchasing a top-of-the-line fiber-optic, multicolor, 1,500-star ceiling—replete with twinkling stars, shooting stars (one every thirty seconds), and a Milky Way, along with whatever background music inspires you. Star ceilings are available for any room of the house, including the bathroom. One version comes with a skylight that gives access to the real night sky. Some companies offer a custom-built cosmos for couples—the stars over your four-poster bed exactly as they appeared on the evening of your marriage. Wrote one enthusiastic reviewer: "Amazing! We live in Las Vegas where we can't see any stars. The [star ceiling] was a whole new gazing experience!" (A short trip out to the desert would have offered a view of the real thing.)

Seti I, pharaoh of the nineteenth Egyptian dynasty (twelfth century BCE), and I appear to have shared a desire to sleep under the stars. The cenotaph, or symbolic tomb, in his mortuary temple in Abydos contains a star ceiling. The scene depicts the body of Nut, the sky goddess, stretched from tiptoes to fingertips. Shu, the god of space, holds her up. There's a ver-

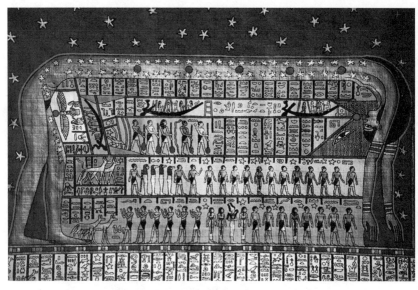

Popular rendition of an Egyptian star ceiling showing the
star-studded body of Sky Goddess Nut stretched across the sky.
(World History Archive/Alamy Stock Photo)

itable astronomy lesson in the accompanying text. It tells when
to anticipate the new year by clocking the movement of Sirius,
or Sothis, the fertility goddess. Her appearance at dawn in
midsummer heralds the annual Nile inundation, which fertilizes the land. The setting sun touches Nut's lips. His wings are
folded, as the sky goddess devours the setting stars, only to
give birth to them next morning. The sun is depicted a second time, on her tiptoes next to an inscription: "The majesty
of this god comes forth from her hinder parts." Farther up on
Nut's thigh, the message continues: "He opens the thighs of his
mother, Nut." A calendar of decans, or ten-day divisions of the
year used to tabulate the months, runs the length of Nut's figure from breast to groin, accompanied by the hieroglyphic

message: "This all happens in the first month of Akhet, at the time of the rising of Sirius."

What drove Seti and other pharaohs to sleep among the stars? The sphinxlike god Tutu, master of the wandering demons and protector of the skies, appears in many of the accompanying astronomical hieroglyphs, especially those that protect the decans (which here are shown as thirty-six groups of small constellations in the Egyptian zodiac that rise consecutively). In addition to being the domain of the visible working gods, the decans were also populated by the invisible demons. Tutu was there to exert a restraining influence on their actions. He protects the king who sleeps eternally beneath him in the afterworld.

Erecting star ceilings in Egyptian temples became especially fashionable during the (Roman) Ptolemaic period (first century BCE). A scene reminiscent of the sort of zodiac familiar to Western stargazers is painted on the ceiling of a chapel on the roof of the Temple of Hathor, the goddess of love and heaven, at Dendera. The constellation map is circular, and displays a crab constellation, a scorpion lady (matching Scorpius), a lion whose tail is held by a goddess (in the location of Leo), and Isis holding an ear of corn (a possible reference to Virgo). There is also a bull's foreleg and a hippo deity. Taweret, the hippo, is the protective goddess of childbirth. She stands on her hind legs and exhibits feminine characteristics, such as pendulous breasts. Egyptians admired the female hippo, who unfailingly protected her young, as opposed to the aggressive, often unpredictable male. One funerary text contains a spell announcing that as the deceased king ascends to heaven, he will be nourished by the dazzlingly white, sweet milk of Taweret.

Contemporary church ceilings also offer inspiration for celestial portraits. In 2004, when workers in Lunenburg, Nova Scotia, began repairing the historic Saint John's Anglican Church, they uncovered hundreds of painted stars on its ceiling. Canadian astronomer David Turner recognized that the pattern wasn't random. Examining old photos, he was able to identify, among others, the constellation of Perseus, positioned high in the east. What struck him as odd, however, was that in the latitude of Nova Scotia, Perseus barely skims the northern horizon. So he backed up the clock and discovered that two thousand years ago the ceiling pattern offered a perfect match to the arrangement on high. The astronomer concluded that the intent of the designers of the star ceiling was to show how the sky over Lunenburg would have appeared during evening twilight on Christmas Eve when Jesus was born.

Star ceilings in public places include the famous painting in the terminal of Grand Central Station in New York City, executed in 1913 by the artist Charles Basing and supervised by Columbia University astronomer Harold Jacoby. It always bothered me because of the incorrect placement of certain constellations. But I was not the first to recognize that something was awry. In 1913, a commuter and amateur astronomer from New Rochelle noticed that the celestial figures were all painted backward. Somewhat angry about having been nearly misled, he noted that Aquarius should be in Pegasus's position, and that Cancer is where Orion ought to be. Professor Jacoby claimed that Basing accidentally placed the model diagram upside down when he laid it at his feet as a painting reference. Jacoby suggested that the artist should have held the diagram over his head while painting rather than transferring an image on the floor to the ceiling. The *New York Times* reported that Basing was unfazed by the finger pointing, and "held that it was a pretty good ceiling for all that." It has also been suggested that the mix-up originated with Jacoby's model, because medieval manuscripts conventionally showed the heavens as seen from the outside.

Stars hold the sky together, and the stars on our rock ceilings hold the rocks together—at least according to the Navajo builders of their star ceilings. There are some fifty of these masterpieces spread across the canyons of northwestern New Mexico and northeastern Arizona. Positioned on rock overhangs and deep alcoves—"planetarium sites," the archaeologists call them—the stars are straight-edged crosses or petal-leaf forms colored orange-red, red, dark gray, blue, and black. They were probably applied by some sort of stamping technique employing leather or cut yucca leaves mounted on long wooden poles and dipped in paint. Actual constellation patterns haven't yet been identified, in part because the Navajo keep them secret. Some interpreters think they might represent stars in general, or maybe specific stars or constellations of special importance.

Why did the Navajo erect star ceilings? Stargazing is a method of divination they use to diagnose illness, discover the sources of ill fortune, and conjure up images of faraway places. For example, in the Great Star Chant the diviner acquires ritual knowledge from the stars. He begins by making a nighttime sand painting of constellations on the floor beneath the stars. Like the ones on the ceiling, the stars are rendered in different colors, and the constellations are said to be spiritually empowered by the particular starshine beneath which they are constructed. The purpose of the sand painting is to attract the healing power of the supernaturals, use it to identify the patient, and absorb the illness from the afflicted one. They say, too, that

the supernaturals are flattered when they see their sand portraits, so they come down from on high to be at one with their
images. Among the constellations identified in sand paintings
are the two *Nahookos* (those who revolve around the central
fire), the Big Dipper and Cassiopeia, who turn around the
hearth of the hogan.

There is no doubt about the identification of sky objects
painted on the Luiseño star ceiling located in southern California's Temecula Valley. A cross-hatched white strip representing the Milky Way extends from one end of the natural
rock shelter to the other. Red disks painted at either extremity
may represent the sun, which crosses the Milky Way when it
reaches the solstice. A Luiseño oral tale recounts the capture
and subsequent escape of the sun from the Milky Way net. The
sun gets caught in the net when it passes through the luminous star clouds of Sagittarius on its way north over the zodiac. When it gets to a northern pocket in the net, it turns and
goes back south. By the time of the equinox it has escaped the
net, only to be recaptured six months later, when it approaches
the winter end of the net at the Milky Way's intersection with
Gemini.

A coming-of-age ceremony for young boys corresponds
with the sun's first capture. It begins as the boys' mothers weave
a net out of milkweed fibers, representing the *wanawat,* or sacred net, that is the Milky Way. Next the men dig a huge trench
in the shape of a human figure and stretch the net across it.
They then bring three large stones from the seashore and space
them equally in the net. An initiate steps onto one of the end
stones, crouches down, then hops from stone to stone. If successful, he eventually bounces high enough to escape the net.
Those who fail, the Luiseño say, will not be long for this world.
The ritual, together with the net in the sky, reminds us that we

must be content knowing we will exist here on earth only for a little while, for our spirits are destined to reside in the Milky Way, our home in the sky.

The Desana are a tribe of Tukanoan people of the equatorial rainforests of the northwest Amazon. They tell the story of their founding hero, a deity who wanders the world in search of a place where his staff, held upright, will cast no shadow. Eventually he locates the divine spot on the equator and there establishes his people. One of the images in the story is a vertical shaft of light that penetrates a womblike lake and fertilizes the earth. This special place is the point of contact of the sky with the earth, and from there life on earth grows in a bounded space around the center. The Desana believe that the hexagon is the fundamental shape that orders all they experience: the honeycomb, the wasp's nest, the plates on the backs of tortoises, and the rock crystals used by shamans to acquire their power. It is imagined to represent the shape of the female womb, as well as the segments that divide the human brain.

The Desana build their longhouses in the shape of a hexagon, and following the hexagonal motif, divide their society into six kinship groups. This model, eternally present and filled with the energy of transformation, offers the Desana a sense of unity in all things. On the natural star ceiling above the Desana village lies the bounded space of the hexagon. It takes on the form of a gigantic constellation that occupies the whole sky. Its vertices are marked by the stars Pollux, Procyon, Canopus, Achernar, Tau-3 in Eridanus, and Capella. The center point is the middle star in Orion's Belt, Alnilam.

The Desana project the immense hexagon on the sky ceiling down to the floor of the forest below, where it marks out the limits of the various Tukanoan tribes. Think of it as a single enormous and transparent hexagonal crystal that stands upright, mapping out the six named waterfalls located on the land's major rivers. The central axis, descended from Alnilam, points to a large boulder covered with petroglyphs, located at an intersection of one of the rivers with the equator.

The Desana hexagon transcends social organization and practical affairs. There's a philosophical and spiritual dimension to how the Desana think about the gigantic constellation on the roof over their heads. The six borders that bound this sacred space make up what they call "The Way," or the path each person must travel through life—a process whereby, with the help of shamanistic counseling and ritual, you become your own self. If you're a male you start at Aldebaran, the place where you're born. You move in a counterclockwise direction to Capella, where you enter the ritual of existing, which means you acquire a name; next you go to Pollux, where initiation into the larger social family takes place; then to Sirius, the star of marriage. Next you turn sharply westward, arriving at the center at Alnilam, the procreation star. Your journey ends where it began, at Aldebaran, the point of death, rebirth, and return to the cosmos. Women begin by journeying in the opposite direction: they marry at Sirius, then they go on jointly with their husbands. For both sexes the course of life, reflected in the cosmic hexagon, becomes a spiritual cycle: from youth to maturity to senescence and back again. Thus the Desana star ceiling is a blueprint for everything that happens on earth—a map that informs biological, cultural, and psychological behavior. The hexagon above guarantees security here below.

Arctic anthropologists recently documented a whole-sky constellation identified for them by Gwich'in shamans in central Alaska, who live in latitude 67 degrees north. "*Yahdii* covers all the skies," one elder told them—more than 143 degrees of it made up mostly of circumpolar stars—the ones that never

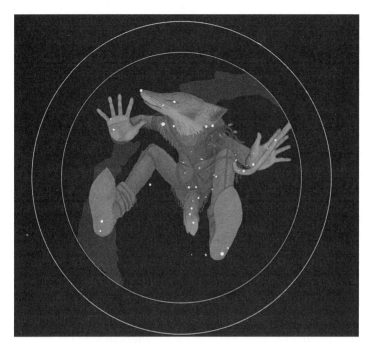

Alaskan Gwich'in Yahdii Constellation. (Illustration by
Mareca Guthrie in consultation with Chris Cannon,
Paul Herbert, and other anonymous Gwich'in elders)

set. The figure is that of a man with a tail. He is crouched face
down and he holds a cane or walking stick in one hand as he
moves slowly from east to west across the night sky. "Yahdii is
walking," says one informant.

Yahdii's left hand is marked by Regulus and surrounding
stars in Leo, his right hand, on the other side of the sky, by
stars in Andromeda. Yahdii's huge coyote-like face is outlined
by his ears, Castor and Capella. His nose is the Pleiades. Specific stars in his legs have yet to be identified, but his splayed-
out feet are indicated by Arcturus and Deneb. Other stars in

Leo, near Regulus, outline the walking stick that Yahdii holds in his hand. His Big Dipper tail drags along the Milky Way, the path he lays down as he journeys over hardpacked snow. Unfortunately there isn't much of a story to tell about the giant man in the sky, because anthropologists are still exploring precisely who Yahdii is; but people assure those who ask after him that he is "up there to watch over us," a perpetual guarantee given the circumpolar location of Yahdii's body parts.

The Gwich'in are related by language to the Dene tribes. Their land-based traveler deity, Yedariye, is characterized as a benevolent spirit "who lives on high." Sometimes he intervenes to help people in distress. Tsimshian natives of the northwestern coastal region of the United States, of whom only 275 native speakers remain, believe their Sky Chief is the keeper of daylight. Once when the world lay in perpetual darkness, clever Raven set out to steal the chief's daylight, a prized possession he kept hidden away in a box in his Upper World house. Raven flew to the east, crashed through the doorway into the glaring daylight of the Upper World, and lit on a branch outside the chief's residence. When the chief's daughter came outside to fetch a bucket of water, Raven transformed into a leaf and fell into the bucket. The young girl drank the water and soon became pregnant. She delivered a bright-eyed baby boy, and as he grew, the boy wailed constantly to play with the daylight box. When the chief finally relinquished the box, the boy transformed back into Raven and escaped with it. Attempting to reenter the world below, and having not eaten for a long time, Raven tried to bargain with the first people: "I'll trade you a little bit of light for some food." When the people refused, in a fit of anger Raven smashed the daylight box, unleashing a flood of light that wiped out the first people.

The Dene conceive of half-human, half-animal creatures who once circled the earth to get rid of dangerous animals or transform them into less threatening forms that roam the land today. The whole-sky constellation of Yahdii may be a cosmic parallel of this supernatural man-animal traveler, the place where he ended his travels when he became an old man and his work was finished. His perpetual appearance in different positions in the night sky provides an excellent backdrop for telling stories about taboos, laws, and other customs, including how to treat animals. But above all, they say, he continues to watch over us, ever vigilant to the possibility that we will stray from our ancient teachings. Then dangerous animals will once again pose a threat to the world.

The early twentieth-century ethnologist John Harrington called the Zuni Chief of the Night

> the most majestic constellation ever reported from any tribe of Indians, as far as we know. It is a gigantic human figure, even bigger than the whole visible sky, for its head had already set in the west or had been obscured by clouds lying near the horizon, while its heart was pointed out in mid heaven, lying in the milky way, and its legs were said to extend beyond the horizon in the east.

Using Harrington's meticulous notes, scholars have identified the heart of the Chief of the Night as corresponding to Vega, Deneb, and part of the Northern Cross. His right arm is Orion's Belt, and a long flexed leg extends from Arcturus in the north to Spica and Antares in the south. Why must he be so large? Zuni elders explain that some part of his body must always be visible because his duty is to guard the earth throughout the year. Sky deities also must guard day and night, always keeping them in equilibrium. It must never be too dark or too light.

Modern star maps display nothing like the Egyptian, Desana, Gwich'in, and Zuni mega-constellations. In the Western tradition, before Cetus reigned the largest was Argo, the ship Jason sailed in search of the Golden Fleece. It sails the southern Milky Way, nearing the southern horizon in winter. Unfortunately we can no longer count it as a single constellation because in 1922 the International Astronomical Union broke it up into component parts: Carina (the keel), Puppis (the stern), Pyxis (the mast), and Vela (the sail). The ship never had a bow. Was it lost in the mist?

10

Gendering the Sky

Iroquois call the sky heralds who precede the seasons runners, constellations that emerge from the horizon and ascend into the night sky. The Seven Sisters are the harbingers of midwinter, while the Seven Brothers bring signs of midsummer. Iroquois storytellers say that during the First Epoch the Seven Sisters looked down from the star-studded sky onto Turtle Island, which had been created out of the back of a giant tortoise. Sadly they saw people gloomily wandering about. "Nobody's smiling," said one sister. "Nobody's dancing," said another. "How can this be?" So they decided to go down to earth and teach the people to sing and dance. A member of the Sky Council warned them, "You know, falling stars frighten those people. The last one fell on a Mohawk village and injured some of them." But still the sisters went. Watching them drop down from heaven, the tribes became frightened by the humming sounds and sparkling light accompanying their descent. Yet as they listened more closely, the people became enchanted by the displays of light and sound, and began imitating what they heard and saw. Soon they found themselves dancing and singing—and smiling. Unfortunately the sisters were sent back into the sky by the councilors as punishment for giving away too much good medicine. Still, we know they did the right thing, for every time they appear we can hear them calling out, "Wake up, Mother Earth! Sing! Dance!"

What about the male complement to the Seven Sisters' story? We know the tale of the Seven Brothers, although their constellation has only recently been identified. Seven young boys were playing in the forest at the edge of the longhouse. Feeling hungry, they went to the clan mother and asked for food, but she was too busy doing chores and refused them. The boys tried to ignore their hunger pangs by returning to play, but they soon needed to beg for nourishment a second time. Again they were shooed away as nuisances. After it happened a third time, the eldest boy made a drum and his brothers began to dance about a sacred tree to the rhythm he tapped out. Soon their feet started to lift off the ground and they rose up to the heavens. All the commotion attracted the clan mother, who looked out of the house to see the boys ascending. "What have I done?" she wailed, realizing that her refusal had driven them off. Hastily she ran inside, grabbed armloads of goodies, and came back outside, beckoning them to return. But too late—the boys were well out of range on the way to Sky World. You can still see them clustered together there. The moral of the story? Keep your food ready, Mother. That's why, to this day, no matter when you arrive, you'll always see a pot of *sagamité* (corn stew) on the fire in the Iroquois longhouse.

Where exactly are the Seven Brothers in the sky? For a long time, ethnographers thought that the Pleiades was used to tell the stories of both the Seven Sisters and the Seven Brothers; but now many think it far more likely that the seven "male" stars lie exactly opposite in the sky to the Pleiades, on the other side of, and equidistant from, the north celestial pole. Follow that line and you'll land among the seven stars of Corona Borealis, called the Council of Chiefs by many North American tribes. (They are easily identified as the cluster at the center of the famous buckskin Pawnee star map.) This constellation, which stands highest just after the start of summer, exactly parallels the Pleiades—creating a perfect mirror image, like everything else in the Iroquois world, and gendered for balance.

You may notice that this Iroquois story, like many that come from other cultures, is quite different from Greco-Roman tales about the origins and names of the constellations. Male deities emerge as the positive characters in Western tales, while women are relegated to secondary, often negative, roles. Take the tale of Callisto: ravished by Zeus—ruler of just about everything and doer of whatever pleases him—she gets turned into a sky bear as punishment for her supposed wrongdoing. Or the story of poor Andromeda, whose beauty leads to her being chained to a rock, desperately in need of Perseus's heroic rescue. A few mythical female figures manage to escape such malign characterizations in Greek fare; for example, angel-winged Virgo the Virgin, goddess of justice, and Atlas's daughters (the Pleiades). Women's accoutrements—like Corona Borealis, the crown of Ariadne; and Coma Berenices, or Berenice's hair, the blond curls of the real King Ptolemy III's wife (which she needed to sacrifice to have her husband returned safely from battle)—are also presented favorably.

Our Greco-Roman ancestors were not the only myth-makers who sound misogynistic. Recall the tale of the Aztec god of sun and war, Huitzilopochtli, who emerges fully clad in armor from his mother's womb and dismembers his treacherous sister, the moon goddess, for plotting the murder of their mother. In China you need to search hard to find the handful of female constellations among the court of the Purple Palace.

One is the "Female Clerk," a single faint star (Psi in Draco) who assists the queen in performing her ritual duties. Ancient palace records say that when her star shines brightly, all recording clerks will tend to use honest language; but when she is faint, they will lean toward dishonesty. The deity Xuanyuan, made up of the Greco-Roman constellation of Leo, is a female rain dragon of fertility. Niu Yu Kung, composed of stars in Ursa Major, managed to achieve the lofty title of the "Inner Consort, Most Exalted of the Seven Ardors."

All three societies—Greco-Roman, Aztec, and Chinese —share a strongly patriarchal history. If there ever were stories told by women, they have not survived. To find a genuine expression of the seminal role of women in society, we need to turn to cultures that are less hierarchically organized, such as the Iroquois, Navajo, and Lakota tribes of North America, and the Aboriginal peoples of Australia, Africa, and the Amazon. In most cases traditional stories told through patterns in the sky, unless held in deep secrecy, have been passed down via both men and women storytellers, and many takeaways from these tales still have practical value.

Gender has always been a basic principle of organization in Iroquois society. Female and male create a balanced astral mythology, each offering narratives that complement the other's. For example, in the Iroquois creation story, Sky Woman and Lynx, the feminine half of society, were responsible for setting in place all cultivable vegetation, while their male counterparts, Our Guardians, made the surrounding trees and wild forest. Epochs of history were gendered as well. Two cycles in the First Epoch of Creation were allocated to women, and one to men. But in the Second Epoch, the two-to-one ratio was flipped. Today we live in the Third Epoch, the first half of which was a male cycle. This epoch won't be complete until the women's cycle emerges to balance it out. It all plays out in the sky. Seasonally, the midsummer constellations are male and the midwinter female. The year happens in paired matching halves. Men convene the Grand Council in summertime and women perform their ritual activities to rouse their "Three Sisters" (corn, beans, and squash) from their long winter sleep. And of course we know of the balance between the seven brothers and seven sisters.

In the mid-1800s, ethnologist Alice Fletcher undertook a dangerous and pioneering cross-country trip to the land of the Sioux. There she collaborated with Native American storytellers and authors and wrote extensively on American Indian education, music, and problems associated with the assimilation of native and European cultures. She also helped write and pass the Dawes Act of 1887, which would ultimately lead to the dissolution of the so-called Indian reservations by allotting tax-free land to individual native people. Working with the Women's National Indian Association, she devoted a lot of attention to women's issues. Alice Fletcher was also the first outsider to tap into the wellspring of cosmically based narratives about the relationship between Native American men and women. Thanks to her largely unheralded efforts, today we possess a wealth of detail about many sky stories.

In one of her earliest treatises about the idea of gender balance that was so fundamental to the native people of the Great Plains, Fletcher wrote of the Pawnee: "Everything is either male or female: these two principles were necessary to the perpetuation of all things." For example, east and south are male; west and north, female; above is male and below is female—just like the Iroquois. Each gender had shrines in the village positioned below their guiding stars (the particular sky identifications have since been lost). There were also stars appearing at the intercardinal directions that mediated between the cardinal shrines. Leadership between males and females alternated in rituals that moved celebrants to stations around the horizon, beginning in the west and proceeding clockwise to the north, east, and south.

As close as she became to her subjects, Alice Fletcher was perhaps too much an outsider, and came along too early, to acquire the rich details behind the dualistic symbolic forces connecting Native American earth and sky. But later investigators, many of mixed blood and living among the native people, have discovered narratives about the relevance of the constellations to relationships between women and men, especially the importance of marriage and family life. One widespread set of stories deals with Star Husbands and Earth Wives. With eighty-six documented versions of the "Star Husband" story, what the story means depends a lot on who gets to tell it.

Two girls camping out gaze at a star-studded sky. Together they pick out a pair of stars, and each makes a wish to be married to her favorite. When they awake, the two find themselves in the Upper World, where they realize that one of the girls is married to a young man star; the other to an old man star. They break the rules to dig an escape hole in the sky, through which they see their home in the Lower World. Longing to return, the girls secure a rope and climb down.

A later version of the Star Husband story, likely influenced by outside scientific knowledge, says that one girl, attracted by a dim star, became the wife of a brave young chief. The star that attracted her was very far away, which accounted for its faintness. By contrast, the other girl had been attracted by a bright star, which only seemed so luminous because it was nearer the earth; she ends up as the wife of a servant. When the servant girl learned that her sister had become the wife of a chief, she wept inconsolably. Still, the girls remained close. They would meet in the clouds where they gathered wild turnips, though the chief's wife, not being required to do any work, would ask her sister to perform the task. One day the servant sister broke an Upper World agricultural rule by twice striking at the base of a turnip, making a hole in the sky. The girls looked down through the hole and saw their home. Desiring to return, they attempted to escape; however, the servant husband went to the chief, who decided to keep them both captive.

Meanwhile, back home, the villagers mourned the disappearance of the girls. One day a group of boys at play looked up and saw two stars coming down from the sky. They ran to the spot where they witnessed the two girls landing. The girls had ropes tied about their waists. The chief had relented. Gathering all the lariats in the upper world, he had knotted them together to make two long ropes. He then had tied one end to each of their waists and gently lowered them back home. In great joy the onlookers carried the news of the sisters' recovery to their sorrowful parents. They told them one of the girls looked sad, the other happy.

In a third version of the story, the girls get home on their own, thanks to the help of a wise old woman of the Upper World who advises them to save the roots of the harvested turnips, knot them together into long ropes, and make their descent. Clearly, the lesson learned from the Star Husband star stories, as in many of our own narratives, has much to do with the storyteller.

Lakota midwives are regarded as shamans who can access the spirits of the stars; they say they are called to their work through dreams. Once labor begins, a midwife is summoned.

She brings special plants to curtail hemorrhaging or to speed up the arrival of the placenta. She prays to Birth Woman (or Blue Woman), who emerges from a hole in the sky located in the bowl of the Big Dipper. Birth Woman guides the baby's spirit into this world and eases pain during the delivery process. She reincarnates spirits in the material world. "And then, after death, she aids those same spirits in their passage out of the material world back through the opening in the Dipper into the spirit world, their place of origin." They say that pregnancy makes a woman stronger. The new life within her adds to her strength. So best not remain idle during that time. The midwife's last duty is to preserve the baby's *chekpa*, "the navel" (actually the umbilical cord). They place it in a beaded pouch which has either a turtle or salamander shape. The turtle shape, associated with girls, conveys steadfastness, long life, and fortitude. Boys instead acquire the salamander properties of adaptability and agility. (Salamanders that lose their tails, for instance, can grow new ones.)

A pair of Lakota constellations symbolize what happens once the baby emerges. The stars of the Square of Pegasus make up the four legs of the Turtle, while the constellation of Cygnus and adjacent stars to the east and west comprise the Salamander. The new mother must pray to the appropriate constellation, asking it to bestow its essence on her newborn. As one storyteller puts it: "The cord between the mother and child is broken at birth, but the cord between the spirit world and children, that connection must be established and then never broken."

Half a world away, the story of the Cowherd and the Weaver Maid is another astral tale that has also changed with time and circumstance. This third-century CE Han dynasty story of a marriage gone wrong casts one of the brightest stars

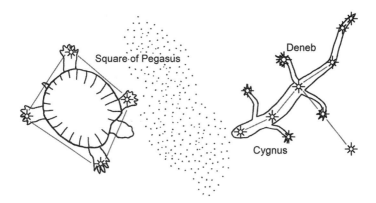

Lakota Turtle and Salamander constellations.
(Sinte Gleska University, Mission, SD, redrawn by Julia Meyerson)

in the summer sky in the female lead role. It begins with young, hard-working Vega, who spends practically all her time laboring at the loom. Her preoccupation worries her father, the Sun King, who would prefer to see his offspring develop other interests—including marriage. He introduces her to the young man Kuo Han (Altair), who tends cattle on the other side of the Silver Stream of Heaven (the Milky Way). The pair immediately fall in love and marry. But then Vega's behavior changes radically. She totally deserts her loom and takes to partying. Dad blames her husband and sets out to separate the couple. He orders the young Cowherd back to his (eastern) side of the river and decrees that the husband and wife will meet only once a year—on the seventh night of the seventh month. He orders a flock of magpies to assemble at the appropriate time to make a bird bridge over the river of stars to support her tiny feet on her annual route across to the eastern side of the Milky Way.

Back goes Vega to her shuttle and Altair to his oxen. When the eagerly awaited time appointed for their first reunion rolls around, the magpies duly assemble, joining their wings so that Weaver Maid can make her crossing and fall into the waiting arms of her joyful lover. In this way the couple continue their marriage for the rest of their lives, though unfortunately when it rains on the day scheduled for the crossing and the River of Heaven becomes flooded, the reunion for that year is canceled. They say that women, the principal worshippers of the marriage of Vega and Altair, hope for skill in the art of weaving and needlework—and they always pray for clear weather on the seventh night of the seventh month. Chinese and Japanese love poetry draw explicitly on the theme of the two cosmic lovers' struggle to cope with prolonged separation:

> Cowherd: When we were separated, I had seen her for a moment only,—And dimly as one sees a flying midge; now I must vainly long for her as before, until the time of our next meeting.

> Weaver Maid: Though for a myriad ages we should remain hand-in-hand and face to face, our exceeding love could never come to an end. (Why then should Heaven deem it necessary to part us?)

Three centuries later, during the Tang dynasty (600–900 CE), a variation of the myth of the Cowherd and the Weaver Maid adds other details about their separation. During the lengthy interval when Cowherd does not lie in conjugal bliss with Weaver Maid, they say that he leaves his cattle in a pen (located in the constellation Capricornus to the south of Aquila) and goes off to make love to Minx Woman, the temptress, at her faraway lunar station. Vega is unfaithful as well. She descends to earth and carries on a series of affairs with the mortal Kuo Han.

In southwestern Africa, the Khoikhoi (Hottentot) "Orion Myth," also called the "Curse of the Women Myth," heralds the power of women. As told to a nineteenth-century English missionary, it begins with the !Khunuseti (the Pleiades), who command their husband: "Go thou and shoot those three zebras (the three stars of Orion's Belt) for us; but if thou dost not shoot, thou darest not come home." Unfortunately their husband dispatched for the hunt carrying only a single arrow. Spotting the zebras, he hastily shot at them with his bow and missed. Meanwhile, a lion stood and looked on from the other side in amusement at the incompetent hunter. The man failed because he didn't have time to retrieve the arrow and set up for another shot. Having been cursed by his wives, the poor hunter was forced to stay out in the cold all night, shivering and suffering from thirst and hunger. The !Khunuseti turned to the other men in the camp and said, "Ye men, do you think you can compare yourselves to us [women], and be our equals? There now, we won't let our own husband come home because he has not killed game." (Incidentally, the missionary pronounced the sky myth as one among a host of insipid and repulsive stories babbled over and over again from the mouths of simple people who had no real understanding of the world.)

Feminine portrayals in constellation stories from the indigenous Boorong of northwest Victoria, Australia, seem much more earthbound than roles familiar in Western star stories. The bird constellations of Yerredetkurrk and Neilloan are among several examples. Yerredetkurrk, the nocturnal fairy owl, lives in the sky, just across the south celestial pole from the Southern Cross. Achernar is her eye. She skims across the treetops at the south horizon, carrying her prey in her talons back to her young ones in the nest. You can hear the *yerr* sound she makes when she flies close by. They call Yerredetkurrk the "mother of wives," the one who establishes behavioral rules of the mother-in-law, son-in-law relationship—rules for setting up appropriate marriages that prevent incestuous relationships. As one anthropologist noted, such rules are a genealogical necessity among small communities like the Boorong. Feminine qualities, such as beauty (she exhibits soft blue-white plumage), mothering, protecting and feeding her young, and maintaining a clean household, make her a totem for southeast Australian women.

Neilloan, the constellation of the malleefowl, a relative of our chicken, resides in the northern sky. Vega marks one of its feet as a reminder of the bird's powerful kick when it constructs a huge mound for its eggs. They say that the Lyrid meteor shower, which radiates from that constellation in April, represents the dirt and twigs that fly through the air during nest building. Because male and female malleefowl mate for life and exhibit strongly defined gender roles, Neilloan reinforces the kind of family life people strive for here on earth, especially cooperation in raising children: males build the nest and tend to the eggs, while females perform feeding duties. (Once the chicks emerge they seem to have little contact with the parents, perhaps because the young ones need to become independent at an early age.)

Interestingly, a number of Boorong bird constellations extol species that mate for life; for example, crows, who find representation in Boorong skies in the constellation named War, whose principal star is Canopus. Likewise, Warepil is the wedge-tailed eagle, Australia's largest bird. Its head is marked by Sirius. He flies in tandem with his wife, Collowgulloric Warepil, which is based in a complex of stars centered around Rigel in southern Orion. Together they soar close to the horizon at sunrise and sunset, surveying their territory. The Boorong say that the proper marriage between the Warepil and Collowgulloric Warepil happens when mates come from opposite moieties, or social groups.

Many indigenous constellations that stress complementary gender relations are accompanied by a kind of down-to-earth, social playfulness in their messaging. My favorite is Ntshune (Fomalhaut in the Greco-Roman Piscis Austrinus). The Tswana (from Botswana) call it the "Kissing Star" because it appears on winter mornings—in time for lovers to part before they are discovered by their parents.

EPILOGUE

In his widely read *Cosmos,* Carl Sagan regards all myths about the sky and the creation of the world as naive because they imagine a universe that follows human or animal precedent. In his view, sky stories differ from our Big Bang scientific idea of a universe that exists for its own sake and that can be observed in order to test—and if necessary, revise—our interpretations of what's going on. Similarly, the astronomer Wayne Orchiston, writing on the astronomy of precolonial Australian people, concludes that because the Aboriginal people didn't understand "the true nature of the various astronomical objects and phenomena they observed, they developed myths which served them by way of explanation." Even the celebrated interpreter of world mythologies Joseph Campbell suggested that the Aboriginal people confused dream reality with daytime events.

Many scientifically trained scholars believe that our human ancestors, lacking the benefit of our technology and the accumulated wisdom of the ages, simply misread the environment, populated it with needless spirits, and based their childlike interpretations of it on false premises. In most scientifically trained minds, the idea that all nature is endowed with life, that it consists of properties transferable to people, and that every material object acts according to its own will, has little value—no matter how effective star stories might be in shaping the everyday lives of the people who tell them. But I think my colleagues err in attempting to assign to myth a rational label that allows it to be dismissed as illusory or wrong.

In no way do I wish to diminish the meaning and value of analytic, scientific thought; rather I have sought to demonstrate here that other effective vehicles, such as associative thought, have been employed in the shared human desire to make sense of the universe. If we look only at whether the results of the discovery process are correct or incorrect, we run the risk of concluding that any ways of comprehending nature other than our own are worthless. And that reduces the possibility of understanding our own modern scientific interpretations and concepts—including where they came from, and how different they may seem from those wonderfully imaginative stories narrated in faraway places and times, when geographical and astronomical knowledge were inseparable from the social and religious values that give human lives meaning.

Every star story I've retold has been used to explain mysterious natural phenomena and events to people in a reassuring way, by relating them to familiar experiences and ideas. It's a way to appease cosmic anxiety. This is why sky objects have names, souls, and even biological functions: for example, the Inuit call mushrooms "star feces," lichen "Sun's urine"; red stars subsist on liver, white ones on kidney. You can't understand and appreciate the astronomical knowledge of any culture without penetrating the depth of social and religious values that make up the lives of its people. The purpose of telling constellation stories, then, is to map one kind of experience, what we view in the sky, onto another—the beginning and the ending of life, the chase, the rescue, the need to survive, the anticipation of seasonal change, or the necessity of sharing a code of good behavior. Every star story is about *us*.

Bibliographic Essay

All websites are current as of February 15, 2019.

Introduction

Readers interested in types of myths and their meanings will find helpful context in D. Leeming, *Creation Myths of the World*, 2 vols. (New York: ABC-CLIO, 2009); and D. Leeming, *The World of Myth: An Anthology* (Oxford: Oxford University Press, 2009).

Most books about constellations tend to be encyclopedic, often arranged by seasonal appearance and with guides to viewing. I recommend:

Bennett, E. *Stars and Constellations*. New York: Scholastic, 2007.
Driscoll, M. *A Child's Introduction to the Night Sky: The Story of the Stars, Planets, and Constellations—and How You Can Find Them in the Sky*. New York: Black Dog and Leventhal, 2004.
Falkner, D. *The Mythology of the Night Sky*. New York: Springer, 2011.

For children, I suggest:

Hislop, S. *Stories in the Stars: An Atlas of Constellations*. New York: Penguin, 2015.
Kerrod, R. *The Book of Constellations*. London: Gary Allen, 2002.
McDonald, M. *Tales of the Constellations: The Myths and Legends of the Night Sky*. New York: Smithmark, 1996.
Mitton, J. *Zoo in the Sky: A Book of Animal Constellations*. Washington, DC: National Geographic Children's Books, 2009.

Oseid, K. *What We See in the Stars: An Illustrated Tour of the Night Sky.* New York: Random House, 2017.

Rey, H. A. *The Stars.* Boston: Houghton Mifflin Harcourt, 2016.

Ridpath, I. *Star Tales.* Revised and expanded edition. Cambridge: Lutterworth, 2018.

Sasaki, C. *The Constellations: Stars & Stories.* New York: Sterling, 2001.

An excellent brief collection of Western constellation myths appears in T. Condos, *Star Myths of the Greeks and Romans: A Sourcebook* (Grand Rapids, MI: Thames, 1997). In addition, see E. C. Krupp, "Sky Tales and Why We Tell Them," in *Astronomy Across Cultures: The History of Non-Western Astronomy,* ed. H. Selin (Dordrecht: Kluwer, 2000), 1–30. Very few sources deal with constellations in other cultures. I can recommend here J. Staal, *The New Patterns in the Sky: Myths and Legends of the Sun, Moon, Stars, and Planets* (New York: Abrams, 1988), and will cite others in the relevant chapters.

A good source for the way ancient priests viewed the constellations is R. Brown, *Researches into the Origin of the Primitive Constellations of the Greeks, Phoenicians, and Babylonians,* 2 vols. (London: Williams and Norgate, 1900).

1

Orion's Many Faces

Useful material on Orion appears in R. Norris and D. Hamacher, "Djulpan: The Celestial Canoe," July 12, 2011, Australian Indigenous Astronomy, at http://aboriginalastronomy.blogspot.com/2011/07/djulpan-celestial-canoe .html (accessed March 8, 2019); G. Ammarell and A. Lowenhaupt Tsing, "Cultural Production of Skylore in Indonesia," in *Handbook of Archaeoastronomy and Ethnoastronomy,* ed. C. Ruggles (New York: Springer, 2014), 2210; D. Freidel, L. Schele, and J. Parker, *Maya Cosmos: Three Thousand Years on the Shaman's Path* (New York: William Morrow, 1993).

The Chinese story of the sons of the emperor was taken from C. Lianshan, *Chinese Myths and Legends* (Cambridge: Cambridge University Press, 2009), 88; see also A. Birrell, *Chinese Mythology: An Introduction* (Baltimore: Johns Hopkins University Press, 1994).

The most complete set of stories about the One-Legged Man appears in E. Magaña, *Orion y la Mujer Pléyades: Simbolismo Astronómico de los Indios Kaliña de Surinam* (Dordrecht: Foris, 1988). See also E. Magaña, "Tropical Tribal Astronomy: Ethnohistorical and Ethnographic Notes," in *Songs from the Sky: Indigenous Astronomical and Cosmological Traditions of the World,* ed. V. del Chamberlain, J. Carlson, and M. J. Young (Leicester, UK: Ocarina Books, 1996), 244–263; and E. Magaña and F. Jara, "The Carib Sky," *Journal de la Société Américanistes* 68 (1982): 105–132.

The story of the Lakota Hand constellation and its social and religious implications is told in R. Goodman, *Lakota Star Knowledge: Studies in Stellar Theology* (Rosebud, SD: Sinte Gleska University, Rosebud Sioux Reservation, 1992); also see R. Goodman, "On the Necessity of Sacrifice in Lakota Stellar Theology as Seen in 'The Hand' Constellation and the Story of the Chief Who Lost His Arm," in *Earth and Sky: Visions of the Cosmos in Native American Folklore,* ed. R. Williamson and C. Farrer (Albuquerque: University of New Mexico Press, 1992), 215–220.

The Maya 260-day cycle approximates the human gestation period. It is also the product of the number of fingers and toes on the human body (twenty) and the number of layers in heaven (thirteen).

The Maya inscription appears in bookofthrees.com/mayan-culture-the -hearth-stones-of-creation. For the Pleiades quotation, see M. Coe, "Native Astronomy in Mesoamerica," in *Archaeoastronomy in Pre-Columbian America,* ed. A. Aveni (Austin: University of Texas Press, 1975), 22–24.

An excellent overview of South African Orion myths can be found in P. G. Alcock, *Venus Rising: South African Beliefs, Customs, and Observations* (Pietermaritzburg, SA: P.G. Alcock, 2014). Unless otherwise noted, this has been my principal resource for all South African material.

2

One Pleiades Fits All

Technically, the Pleiades are not a constellation, but rather an asterism, or a small star group.

There is far too much Pleiades lore in the world to cover in a single chapter. For a general reference, interested readers might begin by consulting "Pleiades in Folklore and Literature," Wikipedia.org.

The Iroquois children's story of the Pleiades is wonderfully told and illustrated by J. Shenandoah and D. George in *Skywoman: Legends of the Iroquois* (Santa Fe, NM: Clear Light, 1998).

"I cannot explain . . ." is taken from J. Fewkes, "The Tusayan Fire Ceremony," *Proceedings of the Boston Society of Natural History* 26 (1895): 453. "There he said . . ." appeared in B. Haile, *Star Lore Among the Navajo* (Santa Fe, NM: Gannon, 1947), 2. "The holistic . . ." is taken from the Navajo Community College General Catalog of 1987; it is also quoted in T. Griffin-Pierce, "Black God: God of Fire, God of Starlight," in *Songs from the Sky: Indigenous Astronomical and Cosmological Traditions of the World,* ed. V. del Chamberlain, J. Carlson, and M. J. Young (Leicester, UK: Ocarina Books, 1996), 73–79.

The tale from contemporary Aboriginal Australia was taken from the narration in M. Andrews, *The Seven Sisters of the Pleiades: Stories from Around the World* (Melbourne: Spinifex, 2000).

The Hesiod passages were taken from *The Works and Days,* in Hesiod, *The Works and Days, Theogony, The Shield of Herakles,* trans. R. Lattimore (Ann Arbor: University of Michigan Press, 1991).

For more on scientific studies correlating the Pleiades appearance with El Niño, see B. Orlove, J. Chiang, and M. Cane, "Ethnoclimatology in the Andes," *American Scientist* 90 (2002): 428–435.

The Spanish chronicler Bernardino de Sahagún writes about the Pleiades in B. de Sahagún, *Florentine Codex: General History of the Things of New Spain, Book 5,* trans. C. Dibble and A. Anderson, Archaeological Institute of America Monograph 14, pt. 5 (Santa Fe School of American Research, 1957).

3
Zodiacs Around the World

Strictly speaking, there ought to be thirteen constellations in the Western zodiac. The thirteenth sign, Ophiuchus, the Serpent Bearer, has his feet tucked in neatly between Scorpio and Sagittarius.

Among the best sources on the zodiac and astrology are R. Gleadow, *The Origin of the Zodiac* (New York: Atheneum, 1969); I. Ridpath, *Star Tales* (New York: Dover, 2011), see also ianridpath.com; and S. Tester, *A History of Western Astrology* (Wolfeboro, NH: Boydell, 1987), source of the quotation by Cecco d'Ascoli on 133.

The words on the Egyptian astrologer's statue are quoted in O. Neugebauer and E. Parker, *Egyptian Astronomical Texts,* vol. 3: *Decans, Planets, Constellations, and Zodiacs* (Providence, RI: Brown University Press, 1969), 214–215; the Assyrian astrologer's quotation comes from A. Oppenheim, "Divination and Celestial Observation in the Last Assyrian Empire," *Centaurus* 14 (1969): 115. For the quotation by the Babylonian priest, see "Prayer for the Gods of the Night," trans. F. T. Stephens, in J. Pritchard, ed., *Ancient Near Eastern Texts Relating to the Old Testament, with Supplements* (Princeton, NJ: Princeton University Press, 1969), 390–391.

For a full discussion of the Greek horoscopic system underlying Western astrology, see A. Aveni, *Conversing with the Planets: How Science and Myth Invented the Cosmos* (New York: Times Books, 1993), ch. 5. This system made use of information from the local positions and motions of the sun, moon, and planets with respect not only to the twelve houses (30-degree strips of the zodiac beginning at the eastern horizon), but also to similar divisions commencing at the vernal equinox, or "first point of Aries." The first 30-degree sector of the house system was the all-important House of the Ascendant. Other houses included Love and Marriage, Death, Honor, Friends, and so on. Planets residing in that sector at the instant of one's birth were thought to exhibit the most potent influence on such matters throughout one's life. In keeping with the overarching principle of dualistic opposition and hierarchical rule, from emperor to peasant, the planets possessed powers that alternated between good and evil, with each planet (in descending order) having less power than the previous one. Planetary positions in the zodiacal signs carried additional meaning. To make matters more complicated, other components and aspects of nature were also tied to the signs, such as gems, metals, herbs, parts of the body, individual organs, bodily fluids (humors), and types of bodily discharge. All were compartmentalized within a universal taxonomy that served as a kind of toolkit for divining. In the hands of an expertly trained astrologer, all entities that make up the ordered world could be brought to bear on a prediction, whether it dealt with medicine and healing, the conduct of war, or how to manage grief over the loss of a loved one. This system has been passed on to modern times in the watered-down, popular form of the daily horoscope.

The twelve talismanic animals in the Chinese system are not based on the ecliptic, like those of the Western zodiac, but rather on the celestial equator. Nor were they pictured in the sky as constellations; instead they were named stations that described Jupiter's twelve-year circuit among the stars in relation to a sixty-year cycle. Likewise the four directional regions were based on the equator rather than on the ecliptic. The twenty-eight *xiu* were more like the Western zodiac in that they were constellations, though equatorially based.

The oft-quoted rhyme about Ho and Hi appears in S. J. Johnson, *Eclipses, Past and Future; with General Hints for Observing the Heavens,* 2nd ed. (London: Parker, 1889), 8. My other Chinese quotation is from A. Pannekoek, *A History of Astronomy* (New York: Dover, 1961), 88. On long-term Chinese planetary conjunctions, see D. Pankenier, "The Mandate of Heaven," *Archaeology* 51 (1998): 26–34, though Pankenier's strict historical interpretation has been challenged. My principal authority on Chinese constellations has been S. Xiaochun and J. Kistemaker, *The Chinese Sky During the Han: Constellating Stars and Society* (Leiden: Brill, 1997).

For more on Maya zodiacs, see A. Aveni, *Skywatchers of Ancient Mexico,* rev. ed. (Austin: University of Texas Press, 2001), 201–203, as well as V. Bricker and H. Bricker, "Zodiacal References in the Maya Codices," in *The Sky on Mayan Literature,* ed. A. Aveni (Austin: University of Texas Press, 1997), 148–183. Contemporary Maya divinatory practice is described in great detail in B. Tedlock, *Time and the Highland Maya* (Austin: University of Texas Press, 1992), esp. ch. 7.

Finally, "As above, so below," the famous adage of astrology, dates back two thousand years to Greek Alexandria, where the astronomer Ptolemy popularized it.

4
Milky Way Sagas

Strictly speaking, the Galaxy, to which astronomers refer with a capital G, implies the one we live in of the multitude of galaxies (small g) that populate the universe.

The version of the Maori Milky Way myth included here was composed by Haritina Mogosanu and appeared in jodcast.net/nztale.html.

The Maya *Popol Vuh* is narrated in several publications. I recommend D. Tedlock, *Popol Vuh: The Mayan Book of the Dawn of Life* (New York: Simon and Schuster, 1996); and A. Christensen, *Popol Vuh: The Sacred Book of the Maya* (Norman: University of Oklahoma Press, 1996). The Maya story of the Milky Way Paddler Gods is told in detail in D. Freidel, L. Schele, and J. Parker, *Maya Cosmos: Three Thousand Years on the Shaman's Path* (New York: Morrow, 1993). S. Milbrath's *Star Gods on the Maya* (Austin: University of Texas Press, 1900), esp. 40–41 and 285–287, contains an excellent compilation of Maya Milky Way imagery.

The Andean chronicler Bernabé Cobo's quotation on the Milky Way appears in G. Urton's "Animals and Astronomy in the Quechua Universe," *Proceedings of the American Philosophical Society* 125 (2) (1981): 113. On the systematic study of the descendants of the Inca, I have referred throughout to Urton's *At the Crossroads of the Earth and the Sky: An Andean Cosmology* (Austin: University of Texas Press, 1981), and for the creation myth, I have adapted Urton's version. The chronicler's translation appears on page 202 of that text. I am indebted to Prof. Urton for his generous participation in dialogues with me on Andean matters over the years. Regarding the Andean

Ausangate ritual today, plans are under way to develop a "spiritual park" to promote biodiversity by growing and consuming certain crops and working to limit grazing animals and mining.

Abundant material on the Barasana Milky Way constellations comes from the work of anthropologist Stephen Hugh-Jones. See, for example, S. Hugh-Jones, *The Palm and the Pleiades: Initiation and Cosmology in Northwest Amazonia* (Cambridge: Cambridge University Press, 1979); and S. Hugh-Jones, "The Pleiades and Scorpius in Barasana Cosmology," in *Archaeoastronomy and Ethnoastronomy in the American Tropics,* ed. A. Aveni and G. Urton (New York: Annals of the New York Academy of Sciences, 1982), 183–201.

G. Lankford's *Reachable Stars* (Tuscaloosa: University of Alabama Press, 2007), 201–210, is my principal source for the Ojibwa and Cherokee Milky Way myths.

The source of the Tabwa myth is A. Roberts, "Perfecting Cosmology: Harmonies of Land, Lake, Body, and Sky," in *African Cosmos, Stellar Arts,* ed. C. Mullen Creamer (Washington, DC: National Museum of African Art of the Smithsonian Institution, 2012), 185.

An excellent discussion of ancient Chinese beliefs about the Milky Way can be found in E. Schafer, *Pacing the Void: T'ang Approaches to the Stars* (Berkeley: University of California Press, 1977), esp. 257–259.

5
Dark Cloud Constellations of the Milky Way

Among sources for emu star stories from Aboriginal Australia, I can recommend R. Fuller, M. Anderson, R. Norris, and M. Trudgett, "The Emu Sky Knowledge of the Kamilaroi and Euahlayi Peoples," *Journal of Astronomical History and Heritage* 17, no. 2 (2014).

Despite casual mention in the literature of dark cloud constellations in the Southern Hemisphere, it wasn't until the late twentieth century that thorough studies in cultural astronomy were conducted in Peru and Australia. In the 1970s and 1980s my colleague Gary Urton, an anthropologist, lived for two years among the residents of Misminay, a small town near the Inca capital of Cuzco. His seminal work, *At the Crossroads of the Earth and the Sky: An Andean Cosmology* (Austin: University of Texas Press, 1981), which I have followed closely and from which I have drawn material quoted in this chapter, has validated a number of ethnohistorical legends and myths by following them into the present. Andean quotes from the colonial period were derived from F. Salomon and J. Urioste, *The Huarochirí Manuscript: A Testament of Ancient and Colonial Andean Religion* (Austin: University of Texas Press, 1991), 372. The Andean quotation about the ewe suckling is taken from H. Livermore, *Garcilaso de la Vega, Royal Commentaries of the Incas* (Austin: University of Texas Press, 1966), 119.

A host of mythic substitutions between Andean and Amazonian constellation stories appear in P. Roe, "Mythic Substitution and the Stars: Aspects of Shipibo and Quechua Ethnoastronomy Compared," in *Songs from the Sky: Indigenous Astronomical and Cosmological Traditions of the World,* ed. V. del Chamberlain, J. Carlson, and M. J. Young (Leicester, UK: Ocarina Books, 1996), 193–228.

The lengthy quotation about the Desana caterpillars was taken from G. Reichel-Dolmatoff, *The Shaman and the Jaguar: A Study of Narcotic Drugs Among the Indians of Colombia* (Philadelphia: Temple University Press, 1975), 116.

6
Polar Constellations

The Inuit star stories narrated in this chapter and all the quotations connected with them are derived from J. MacDonald, *The Arctic Sky: Inuit Astronomy, Star Lore, and Legend* (Toronto: Royal Ontario Museum and Nunavut Research Institute, 1998). I owe John MacDonald much gratitude for helpful, engaging, and continued discussions over the years.

Northern bear stories are also recorded and analyzed in G. Lankford, *Reachable Stars* (Tuscaloosa: University of Alabama Press, 2007). See also E. C. Krupp, *Beyond the Blue Horizon: Myths of the Sun, Moon, Stars, and Planets* (Oxford: Oxford University Press, 1992), esp. ch. 14. The Kiowa bear story was adapted from N. Scott Momaday, "The Seven Sisters," in *Songs from the Sky: Indigenous Astronomical and Cosmological Traditions of the World,* ed. V. del Chamberlain, J. Carlson, and M. J. Young (Leicester, UK: Ocarina Books, 1996).

T. Condos, *Star Myths of the Greeks and Romans: A Sourcebook* (Grand Rapids, MI: Thames, 1997), gives an excellent narration of the Callisto bear story.

The Great Lakes Fox story about chasing the bear is taken from E. Dempsey, "Aboriginal Canadian Sky Lore of the Big Dipper," *Journal of the Royal Astronomical Society of Canada* 102 (2008): 59–60.

The movement of the celestial pole through time, due to the precession of the equinoxes, is explained in detail in A. Aveni, *Skywatchers of Ancient Mexico,* rev. ed. (Austin: University of Texas Press, 2001), ch. 3.

Associative thinking is discussed in J. Goody, *Domestication of the Savage Mind* (Cambridge: Cambridge University Press, 1977), 40 and 68. Goody regards it as a form of thinking linked with "primitive societies," a misleading designation. Unlike causal thought, associative or coordinative thought sought to systematize events and things into a structured pattern that accounted for the mutual influences of all its parts.

7
Star Patterns in the Tropics

The delightful Maui myth comes from W. Westervelt, *Legends of Maui* (Honolulu: Hawaiian Gazette, 1910); the story of Little Eyes is in R. Craig, *Handbook of Polynesian Mythology* (Santa Barbara, CA: ABC-Clio 2004), 207–208. Instructions on how to make a star compass out of a gourd are reprinted in *Report of the Minister of Public Instruction to the President of the Republic of Hawaii for the Biennial Period ending December 31st 1899* (Honolulu: Hawaiian Gazette Company Print, 1900), 34.

A comparison of tropical and polar systems of sky orientation is elaborated in A. Aveni, "Tropical Archaeoastronomy," *Science* 213 (1981): 161–171, and some of the material on tropical navigation in this chapter is adapted from chapter 3 of A. Aveni, *People and the Sky* (London: Thames and Hudson, 2008). I think the best reference on tropical navigation is D. Lewis, *We the Navigators* (Honolulu: University of Hawaii Press, 1972). Lewis was foremost among the experimental navigators who sailed long distances using the tiborau's techniques.

For the Tahitian sea chant, see "The Birth of New Lands, After the Creation of Havai'i (Raiatea)," *Journal of the Polynesian Society* 3 (1894): 186–189, quotation on 187.

The concept of linear constellations is dealt with in C. Kursh and T. Kreps, "Linear Constellations in Tropical Navigation," *Current Anthropology* 15 (1974): 334–337.

The instructions on Hawaiian navigation using a hollowed-out gourd appear in *Kamakau: Hawaiian Annual, 1891.* They are discussed in detail in B. Penprase, *The Power of Stars* (New York: Springer, 2010), 61.

The Tahitian navigator quotation appears in G. Deming, "The Geographical Knowledge of the Polynesians and the Nature of Inter-Island Contact," *Journal of the Polynesian Society* 71 (1962): 111.

The comments by the German sea captain are documented in Board of Regents, Smithsonian Institution, *Annual Report of the Smithsonian Institution, for the Year Ending 1899*, 488, available at https://books.google.com /books?id=9WiPjla-KEkC&source=gbs_navlinks_s.

For more on tropical constellations, see the section "The Hawaiian and Polynesian Sky," in chapter 2 of B. Penprase, *The Power of Stars: How Celestial Observations Have Shaped Civilization* (New York: Springer, 2011).

8
Empire in the Sky

The Lakota quotation appears in A. Bird, "Astronomical Star Lore of the Lakota Sioux: Lakota Ethnoastronomy," 2012, sccass-international.com.

I think the best resource on the architecture of the Navajo hogan is T. Griffin-Pierce, "The *Hooghan* and the Stars," in *Earth and Sky: Visions of the Cosmos in Native American Folklore*, ed. R. Williamson and C. Farrer (Albuquerque: University of New Mexico Press, 1992), 110–130. On the Pawnee lodge, see V. del Chamberlain, *When Stars Came Down to Earth: Cosmology of the Skidi Pawnee Indians of North America* (Los Altos, CA: Ballena, 1982). And on the Pawnee village, see A. Fletcher, "Star Cult Among the Pawnee—A Preliminary Report," *American Anthropologist* 4 (1902): 730–736. For the Murie quotation, see J. Murie, "Ceremonies of the Pawnee, Part 1: The Skiri," ed. D. Parks, in *Smithsonian Contributions to Anthropology*, no. 27 (Washington, DC: Smithsonian Institution Press, 1981), 76.

Material on the Kiribati house was drawn from M. Makemson, "Hawaiian Astronomical Concepts," *American Anthropologist* 40, no. 3 (1938): 370–383. Grimble is quoted in R. Grimble, *Migrations, Myth, and Magic from the Gilbert Islands: Early Writings of Sir Arthur Grimble* (London: Routledge, 1972), 229.

More on cosmic aspects of domestic architecture, including an interesting cosmic house analogy from South America, can be found in J. Wilbert, "Warao Cosmology and Yekuana Roundhouse Symbolism," *Journal of Latin American Lore* 7 (1981): 37–72. The Lakota example was drawn from R. Goodman, *Lakota Star Knowledge: Studies in Lakota Stellar Theology* (Rosebud, SD: Sinte Gleska University, Rosebud Sioux Reservation, 1992).

The Pawnee story about the origin of death is taken from F. Boas, "The Origin of Death," *Journal of American Folklore* 30 (1917): 486–491.

Western readers often find Chinese constellations and Chinese urban planning confusing. I think that's because these topics are often presented using an amalgam of stories and images taken from many historical periods. Among my most dependable references is Zhiyi Zhou, n.d., "Suzhou in History: City Layout and Urban Culture," available at https://www.fordham .edu/downloads/file/5697/zhou_-_suzhou_in_history; this is the source of my quoted material. For the most reliable Chinese constellation maps, see S. Xiaochun and J. Kistemaker, *The Chinese Sky During the Han: Constellating Stars and Society* (Leiden: Brill, 1997). Additional star stories from China were drawn from "Legends of Chinese Asterisms," available at the Hong Kong Space Museum website, https://www.lcsd.gov.hk/CE/Museum /Space/archive/StarShine/Starlore/e_starshine_starlore14.htm. The quotations on the alignment of Beijing were taken from P. Wheatley, *The Origins and Character of the Ancient Chinese City*, vol. 2: *The Chinese City in Comparative Perspective* (New Brunswick, NJ: Aldine, 2008), 461; and J. Needham, *Science and Civilization in China*, vol. 3 (Cambridge: Cambridge University Press, 1959), 82.

For more on the plan of Washington, see J. Meyer, *Myths in Stone: The Religious Dimensions of Washington, D.C.* (Berkeley: University of California Press, 2001). The 1792 essay is quoted in M. Baker, "Surveys and Maps, District of Columbia," *National Geographic* 6 (1895): 154.

9
Star Ceilings and Mega-Constellations

An excellent survey of the Poseidon, Perseus, and Andromeda entanglement can be found on shmoop.com/perseus-andromeda.

On the art of installing a star ceiling, see, for example, http://calsworld .net/StarCeilings.htm.

Another Egyptian star ceiling appears on the tomb of the eighteenth-century BCE Egyptian king Senenmut at Thebes. For details see M. Clagett, *Ancient Egyptian Science*, vol. 2 (Philadelphia: American Philosophical Society, 1995). The protective role of Egyptian God Tutu in star ceilings is explained in O. E. Kuper, *The Egyptian God Tutu: A Study of the Sphinx God and Master of Demons with a Corpus of Documents* (Leuven: Peeters, 2003), 67–70. I have followed the interpretation of R. Gleadow, *The Origin of the Zodiac* (New York: Atheneum, 1969) regarding the figure of the Egyptian deity Nut.

The story of the Lunenburg star ceiling is told in D. Falk, "'Ancient' Stars Shine On," available at the *Astronomy* magazine website http://www.astron omy.com/news/2004/10/ancient-stars-shine-on. Star ceilings adorn many contemporary religious structures, and mostly depict the positions of the constellations on the evening of the building's dedication (see, e.g., the Lovely Lane Methodist Church in Baltimore designed by Stanford White).

The amusing story behind the backward ceiling in New York's Grand Central Terminal was taken from Julia Goicochea, "The Story Behind Grand Central Terminal's Beautiful Ceiling," CultureTrip.com, March 28, 2018, available at https://theculturetrip.com/north-america/usa/new-york/articles/the -story-behind-grand-central-terminals-beautiful-ceiling.

Native American material in this chapter was acquired from V. del Chamberlin, "Navajo Indian Star Ceilings," in *World Archaeoastronomy*, ed. A. Aveni (Cambridge: Cambridge University Press, 1989), 331–339; M. J. Young and R. Williamson, "Ethnoastronomy: The Zuni Case," in *Archaeo-astronomy in the Americas*, ed. R. Williamson (Los Altos, CA: Ballena, 1981), 183–192 (in which the Harrington quotation appears, on 187); and T. Griffin-Pierce, "The *Hooghan* and the Stars," in *Earth and Sky: Visions of the Cosmos in Native American Folklore*, ed. R. Williamson and C. Farrer (Albuquerque: University of New Mexico Press, 1992), 110–130. E. C. Krupp, *Beyond the Blue Horizon: Myths of the Sun, Moon, Stars, and Planets* (Oxford: Oxford University Press, 1992), 269–270, describes the Luiseño initiation ceremony.

The story of the Desana hexagon is analyzed in detail in G. Reichel-Dolmatoff, "Astronomical Models of Social Behavior Among Some Indians of Colombia," in *Ethnoastronomy and Archaeoastronomy in the American Tropics*, ed. A. Aveni and G. Urton (New York: Annals of the New York Academy of Sciences, 1982), 165–181.

For the Gwich'in all-sky constellation, see C. Cannon and G. Holton, "A Newly Documented Whole-Sky Circumpolar Constellation," *Arctic Anthro-pology* 51 (2014): 1–8. I am indebted to Chris Cannon for correspondence regarding his ongoing work with the Gwich'in people and for generously allowing me to use the all-sky constellation he has been studying as an illustration.

The Theft of Daylight/Raven myth is analyzed in detail in D. Vogt, "Raven's Universe," in *Songs from the Sky: Indigenous Astronomical and Cosmo-logical Traditions of the World*, ed. V. del Chamberlain, J. Carlson, and M. J. Young (Leicester, UK: Ocarina Books, 1996), 38–48.

10

Gendering the Sky

On creations and Iroquois gendering, my principal source for the sky stories has been B. Mann, *Iroquoian Women: The Gantowisas* (New York:

Peter Lang, 2000). I agree with Mann's identification of Corona Borealis as the male complement of the female Pleiades.

My quotations on Pawnee gender concepts came from A. Fletcher, "Star Cult Among the Pawnee—A Preliminary Report," *American Anthropologist* 4 (1902): 730–736. G. Lankford, *Reachable Stars* (Tuscaloosa: University of Alabama Press, 2007) documents the eighty-six versions of the "Star Husband" story.

Quotations about the Lakota Turtle-Salamander constellations were taken from R. Goodman, *Lakota Star Knowledge: Studies in Lakota Stellar Theology* (Rosebud, SD: Sinte Gleska University, Rosebud Sioux Reservation, 1992).

E. Schafer, *Pacing the Void: T'ang Approaches to the Stars* (Berkeley: University of California Press, 1977), offers the most complete accounts of the multiple versions of the Chinese Cowherd and the Weaver Maid tale. The eighth-century poetry quoted in this chapter was taken from L. Hearn, *The Romance of the Milky Way and Other Stories and Studies* (Boston: Houghton and Mifflin, 1907), 33 and 40.

The "Curse of the Women" quotation was taken from T. Hahn, *Tsuni-Goam: The Supreme Being of the Khoi Khoi* (London: Trubner, 1881).

On the Australian constellations related to gender, see J. Morieson, *Stars over Tyrrell: The Night Sky Legacy of the Boorong* (Victoria: Sea Lake Historical Society, 2000).

Epilogue

For this summation chapter, I benefited especially from conversations with astronomer Ed Krupp and anthropologist John MacDonald, having read and quoted from their books: E. C. Krupp, *Beyond the Blue Horizon: Myths of the Sun, Moon, Stars, and Planets* (Oxford: Oxford University Press, 1992), esp. 16–21; and J. MacDonald, *The Arctic Sky: Inuit Astronomy, Star Lore, and Legend* (Toronto: Royal Ontario Museum and Nunavut Research Institute, 1998), esp. 17–19.

Carl Sagan's interpretation of several indigenous myths of creation comes from *Cosmos* (New York: Random House, 1980), 257–260. The quotation on the Aboriginal people was taken from W. Orchiston, "Australian Aboriginal, Polynesian, and Maori Astronomy," in *Astronomy Before the Telescope*, ed. C. Walker (London: British Museum Press, 1996), 320.

Acknowledgments

Thanks to Yale University Press for another opportunity to work with a professional, skilled group of enabling people, especially senior editor Joe Calamia for suggesting the idea of a book about cross-cultural star stories in casual conversation, then following through with a design plan that greatly enhanced my words. I am grateful for the sharp eyes and acute mindfulness of editorial assistant Michael Deneen, senior manuscript editor Susan Laity, copyeditor Julie Carlson, proofreader Erica Hanson, and indexer Alexa Selph. Joe's plan also included the brilliant idea to enlist artist Matthew Green, who created the artwork that appears at the head of each chapter. Once again I had the opportunity to display the artistic skills of Julia Meyerson, who drew a number of the illustrative constellation maps. Thank you, Matthew and Julia, for images that suit my words so well. And thanks once again as well to my agents Faith Hamlin (has it really been thirty-two years?) and Ed Maxwell at Sanford J. Greenburger Associates.

I thank my colleague, the classicist Robert Garland, for many lengthy discussions on interdisciplinary teaching and writing, and especially for sharing his wisdom on the mythology of the classical world; the same for anthropologist Gary

Urton, particularly on Andean matters. I extend my thanks as well to my colleagues in Native American Studies Chris Vecsey and Carol Ann Lorenz, to John MacDonald for sharing his deep knowledge of Inuit culture, and to astronomer Ed Krupp, who has always been as interested as I in all aspects of cultural astronomy. I am grateful to anthropologist Chris Cannon for sharing information and ideas on Gwich'in constellations.

On the home front I am continually grateful to Joe Eakin, director of Colgate University's Ho Tung Visualization Laboratory, for innovatively bringing star story imagery to life on a big dome for audience presentation and interactive participation. Lastly, I remain continually grateful to my impeccable assistant and veteran collaborator on twelve previous book manuscripts, Diane Janney, and to my spouse, Lorraine Aveni, an unceasingly careful reader and constructive listener.

Index

Page numbers in *italics* indicate illustrations.